BARRON'S

SAT* SUBJECT TEST

MATH LEVEL 2

13TH EDITION

Richard Ku, M.A.

BARRON'S

About the Author

Richard Ku taught secondary mathematics for more than 30 years and has coached math teams for 15 years. He has been an AP Reader in both Calculus and Statistics and has served as an AP Statistics Table Leader.

Acknowledgments

Thanks go to my wife, Doreen, for her love and support during the preparation of this manuscript. Thanks also go to my friend Pat Rebollo, whose insights on how students learn have proven invaluable during the writing of this book.

NOTE TO CUSTOMER

Go to *barronsbooks.com/tp/sat/math2/* or scan the QR code below to access the two SAT Math Level 2 online tests.

*Be sure to have your copy of *SAT Subject Test Math Level 2, 13th Edition* on hand to complete the registration process.

All inquiries should be addressed to:
Barron's Educational Series, Inc.
250 Wireless Boulevard
Hauppauge, New York 11788
www.barronseduc.com

ISBN: 978-1-4380-1114-1

ISSN 1939-3237

10% POST-CONSUMER WASTE
Paper contains a minimum of 10% post-consumer waste (PCW). Paper used in this book was derived from certified, sustainable forestlands.

PRINTED IN THE UNITED STATES OF AMERICA

9 8 7 6 5 4 3 2 1

Contents

PART 1
DIAGNOSTIC TEST

PART 2
REVIEW OF MAJOR TOPICS

**PART 3
PRACTICE TESTS**

Note: The link to the two online practice tests can be found on the inside book cover and on the copyright page. These practice tests can be accessed on most mobile devices, including tablets and smartphones.

Introduction

The purpose of this book is to help you prepare for the SAT Level 2 Mathematics Subject Test. This book can be used as a self-study guide or as a textbook in a test preparation course. It is a self-contained resource for those who want to achieve their best possible score.

Because the SAT Subject Tests cover specific content, they should be taken as soon as possible after completing the necessary course(s). This means that you should register for the Level 2 Mathematics Subject Test in June after you complete a precalculus course.

You can register for SAT Subject Tests at the College Board's website, *www.collegeboard.org*; by calling (866) 756-7346, if you previously registered for an SAT Test or SAT Subject Test; or by completing registration forms in the SAT Registration Booklet, which can be obtained in your high school guidance office. You may register for up to three SAT Subject Tests at each sitting.

> **IMPORTANT REMINDER**
>
> Be sure to check the official College Board website for the most accurate information about how to register for the test and what documentation to bring on test day.

Colleges use SAT Subject Tests to help them make both admission and placement decisions. Because the SAT Subject Tests are not tied to specific curricula, grading procedures, or instructional methods, they provide uniform measures of achievement in various subject areas. This way, colleges can use SAT Subject Test results to compare the achievement of students who come from varying backgrounds and schools.

You can consult college catalogs and websites to determine which, if any, SAT Subject Tests are required as part of an admissions package. Many "competitive" colleges require the Level 1 Mathematics Subject Test.

If you intend to apply for admission to a college program in mathematics, science, or engineering, you may be required to take the Level 2 Mathematics Subject Test. If you have been generally successful in high school mathematics courses and want to showcase your achievement, you may want to take the Level 2 Mathematics Subject Test and send your scores to colleges you are interested in even if it isn't required.

OVERVIEW OF THIS BOOK

A Diagnostic Test in Part 1 follows this introduction. This test will help you quickly identify your weaknesses and gaps in your knowledge of the topics. You should take it under test conditions (in one quiet hour). Use the Answer Key immediately following the test to check your answers, read the explanations for the problems you did not get right, and complete the Self-Evaluation Chart that follows the explanations. These explanations include a code for calculator use, the correct answer choice, and the chapter number of the relevant topic in Part 2 "Review of Major Topics." For your convenience, the Self-Evaluation Chart is also keyed to these locations.

The majority of students who take the Level 2 Mathematics Subject Test are accustomed to using graphing calculators. Where appropriate, explanations of problem solutions are based on their use. Secondary explanations that rely on algebraic techniques may also be given.

Part 3 contains six practice tests. The breakdown of test items by topic approximately reflects the nominal distribution established by the College Board. The percentage of questions for which calculators are required or useful on the practice tests is also approximately the same as that specified by the College Board. Similar to the College Board tests, each practice test in this book opens with a "Reference Information" box that contains a list of four solid geometry formulas. The practice tests are self-contained. Each has an answer sheet and a complete set of directions. Each test is followed by an Answer Key, explanations such as those found after the Diagnostic Test, and a Self-Evaluation Chart.

A summary of formulas is given after Practice Test 6. These formulas are organized by the chapters of the review section.

In addition to the six practice tests in the book, there are also two additional full-length online practice tests that can be accessed on your computer, smartphone, or tablet. Detailed answers and explanations are provided for every question, and these tests automatically calculate your raw and scaled scores.

To access these online tests, go to: *barronsbooks.com/tp/sat/math2/*

OVERVIEW OF THE LEVEL 2 MATHEMATICS SUBJECT TEST

The SAT Level 2 Mathematics Subject Test is one hour in length and consists of 50 multiple-choice questions, each with five answer choices. The test is aimed at students who have had two years of algebra, one year of geometry, and one year of trigonometry and elementary functions. According to the College Board, test items are distributed over topics as follows:

- Number and Operations: 5–7 questions
  Operations, ratio and proportion, complex numbers, counting, elementary number theory, matrices, sequences, series, and vectors

- Algebra and Functions: 24–26 questions
  Work with equations, inequalities, and expressions; know properties of the following classes of functions: linear, polynomial, rational, exponential, logarithmic, trigonometric and inverse trigonometric, periodic, piecewise, recursive, and parametric; be familiar with mathematical modeling techniques

- Coordinate Geometry: 5–7 questions
  Symmetry, transformations, conic sections, polar coordinates

- Three-Dimensional Geometry: 2–3 questions
  Volume and surface area of solids (prisms, cylinders, pyramids, cones, and spheres); coordinates in three dimensions

- Trigonometry: 6–8 questions
  Right triangles; laws of sines and cosines; radian measure; equations, identities, and double angle formulas

- Data Analysis, Statistics, and Probability: 4–6 questions
  Measures of central tendency and spread; graphs and plots; least squares regression (linear, quadratic, and exponential); probability

CALCULATOR USE

NOTE

To make them as specific and succinct as possible, calculator instructions in the answer explanations are based on the TI-83 and TI-84 families of calculators.

As noted earlier, most students who take the Level 2 Mathematics Subject Test are familiar with the use of a graphing calculator. In addition to performing the calculations of a scientific calculator, graphing calculators can be used to analyze graphs and to find zeros, points of intersection of graphs, and maxima and minima of functions. Graphing calculators can also be used to find numerical solutions to equations, generate tables of function values, evaluate statistics, and find regression equations. The author assumes that readers of this book plan to use a graphing calculator when taking the Level 2 Mathematics Subject Test.

You should always read a question carefully and decide on a strategy to answer it before deciding whether a calculator is necessary. A calculator is useful or necessary on only 55–65 percent of the questions. You may find, for example, that you need a calculator only to evaluate some expression that must be determined based solely on your knowledge about how to solve the problem. In both the Diagnostic Test and the Practice Tests, an asterisk (*) in the Answers Explained sections indicates questions for which a graphing calculator is necessary or helpful.

Most graphing calculators are user friendly. They follow the order of operations, and expressions can be entered using several levels of parentheses. There is never a need to round and write down the result of an intermediate calculation and then rekey that value as part of another calculation. Premature rounding can result in choosing a wrong answer if numerical answer choices are close in value.

On the other hand, graphing calculators can be troublesome or even misleading. For example, if you have difficulty finding a useful window for a graph, perhaps there is a better way to solve a problem. Piecewise functions, functions with restricted domains, and functions that have asymptotes are other examples where the usefulness of a graphing calculator may be limited.

Calculators have popularized a multiple-choice problem-solving technique called backsolving, where answer choices are entered into the problem to see which works. In problems where decimal answer choices are rounded, none of the choices may work satisfactorily. Be careful not to overuse this technique.

The College Board has established rules governing the use of calculators on the Mathematics Subject Tests:

- You may bring extra batteries or a backup calculator to the test. If you wish, you may bring both scientific and graphing calculators.
- Test centers are not expected to provide calculators, and test takers may not share calculators.
- Notify the test supervisor to have your score canceled if your calculator malfunctions during the test and you do not have a backup.
- Certain types of devices that have computational power are **not permitted**: cell phones, pocket organizers, powerbooks and portable handheld computers, and electronic writing pads. Calculators that require an electrical outlet, make noise or "talk," or use paper tapes are also prohibited. A list of acceptable calculators can be found on the College Board's website: *collegereadiness.collegeboard.org/sat/taking-the-test/calculator-policy*
- You do not have to clear a graphing calculator's memory before or after taking the test. However, any attempt to take notes in your calculator about a test and remove it from the room will be grounds for dismissal and cancellation of scores.

HOW THE TEST IS SCORED

There are 50 questions on the Level 2 Mathematics Subject Test. Your raw score is the number of correct answers minus one-fourth of the number of incorrect answers, rounded to the nearest whole number. For example, if you get 30 correct answers, 15 incorrect answers, and leave 5 blank, your raw score would be $30 - \frac{1}{4}(15) \approx 26$, rounded to the nearest whole number.

Raw scores are transformed into scaled scores between 200 and 800. The formula for this transformation changes slightly from year to year to reflect varying test difficulty. In recent years, a raw score of 44 was high enough to transform to a scaled score of 800. Each point less in the raw score resulted in approximately 10 points less in the scaled score. For a raw score of 44 or more, the approximate scaled score is 800. For raw scores of 44 or less, the following formula can be used to get an approximate scaled score on the Diagnostic Test and each Practice Test:

$S = 800 - 10(44 - R)$, where S is the approximate scaled score and R is the rounded raw score

The Self-Evaluation page for the Diagnostic Test and each Practice Test includes space for you to calculate your raw score and scaled score.

STRATEGIES TO MAXIMIZE YOUR SCORE

Leave your cell phone at home, in your locker, or in your car!

- **Budget your time.** Although most testing centers have wall clocks, you would be wise to have a watch on your desk. Since there are 50 items on a one-hour test, you have a little over a minute per item. Typically, test items are easier near the beginning of a test, and they get progressively more difficult. Don't linger over difficult questions. Work the problems you are confident about first, and then return to the ones that are difficult for you later.
- **Guess intelligently.** You are likely to get a higher score if you can confidently eliminate two or more answer choices and a lower score if you can't eliminate any choices.

- **Read the questions carefully.** Answer the question asked, not the one you may have expected. For example, you may have to solve an equation to answer the question, but the solution itself may not be the answer.
- **Mark answers clearly and accurately.** Since you may skip questions that are difficult, be sure to mark the correct number on your answer sheet. If you change an answer, erase cleanly and leave no stray marks. Mark only one answer; an item will be graded as incorrect if more than one answer choice is marked.
- **Change an answer only if you have a good reason for doing so.** It is usually not a good idea to change an answer on the basis of a hunch or whim.
- **As you read a problem, think about possible computational shortcuts to obtain the correct answer choice.** Even though calculators simplify the computational process, you may save time by identifying a pattern that leads to a shortcut.
- **Substitute numbers to determine the nature of a relationship.** If a problem contains only variable quantities, it is sometimes helpful to substitute numbers to understand the relationships implied in the problem.
- **Think carefully about whether to use a calculator.** The College Board's guideline is that a calculator is useful or necessary in about 60% of the problems on the Level 2 Test. An appropriate percentage for you may differ from this, depending on your experience with calculators. Even if you learned the material in a highly calculator-active environment, you may discover that a problem can be solved more efficiently without a calculator than with one.
- **Check the answer choices.** If the answer choices are in decimal form, the problem is likely to require the use of a calculator.

STUDY PLANS

Your first step is to take the Diagnostic Test. This should be taken under test conditions: timed, quiet, without interruptions. Identify areas of weakness using the cross-references to the Part 2 review. Use the review to strengthen your understanding of the concepts involved.

Ideally, you would start preparing for the test two to three months in advance. Each week, you would be able to take one Practice Test, following the same procedure as for the Diagnostic Test. Depending on how well you do, it might take you anywhere between 15 minutes and an hour to identify your areas of weakness using the cross-references to the Part 2 review and review the relevant material from Part 2. Obviously, if you have less time to prepare, you would have to intensify your efforts to complete the six Practice Tests, or do fewer of them.

The best way to use Part 2 of this book is as reference material. You should look through this material quickly before you take the practice tests, just to get an idea of the range of topics covered and the level of detail. However, these parts of the book are more effectively used after you've taken and corrected a practice test.

Before your review and practice is complete, be sure to visit *barronsbooks.com/tp/sat/math2/* to take both online practice tests!

PART 1
Diagnostic Test

PART 1
Diagnostic Test

ANSWER SHEET
Diagnostic Test

1. Ⓐ Ⓑ Ⓒ Ⓓ Ⓔ
2. Ⓐ Ⓑ Ⓒ Ⓓ Ⓔ
3. Ⓐ Ⓑ Ⓒ Ⓓ Ⓔ
4. Ⓐ Ⓑ Ⓒ Ⓓ Ⓔ
5. Ⓐ Ⓑ Ⓒ Ⓓ Ⓔ
6. Ⓐ Ⓑ Ⓒ Ⓓ Ⓔ
7. Ⓐ Ⓑ Ⓒ Ⓓ Ⓔ
8. Ⓐ Ⓑ Ⓒ Ⓓ Ⓔ
9. Ⓐ Ⓑ Ⓒ Ⓓ Ⓔ
10. Ⓐ Ⓑ Ⓒ Ⓓ Ⓔ
11. Ⓐ Ⓑ Ⓒ Ⓓ Ⓔ
12. Ⓐ Ⓑ Ⓒ Ⓓ Ⓔ
13. Ⓐ Ⓑ Ⓒ Ⓓ Ⓔ

14. Ⓐ Ⓑ Ⓒ Ⓓ Ⓔ
15. Ⓐ Ⓑ Ⓒ Ⓓ Ⓔ
16. Ⓐ Ⓑ Ⓒ Ⓓ Ⓔ
17. Ⓐ Ⓑ Ⓒ Ⓓ Ⓔ
18. Ⓐ Ⓑ Ⓒ Ⓓ Ⓔ
19. Ⓐ Ⓑ Ⓒ Ⓓ Ⓔ
20. Ⓐ Ⓑ Ⓒ Ⓓ Ⓔ
21. Ⓐ Ⓑ Ⓒ Ⓓ Ⓔ
22. Ⓐ Ⓑ Ⓒ Ⓓ Ⓔ
23. Ⓐ Ⓑ Ⓒ Ⓓ Ⓔ
24. Ⓐ Ⓑ Ⓒ Ⓓ Ⓔ
25. Ⓐ Ⓑ Ⓒ Ⓓ Ⓔ
26. Ⓐ Ⓑ Ⓒ Ⓓ Ⓔ

27. Ⓐ Ⓑ Ⓒ Ⓓ Ⓔ
28. Ⓐ Ⓑ Ⓒ Ⓓ Ⓔ
29. Ⓐ Ⓑ Ⓒ Ⓓ Ⓔ
30. Ⓐ Ⓑ Ⓒ Ⓓ Ⓔ
31. Ⓐ Ⓑ Ⓒ Ⓓ Ⓔ
32. Ⓐ Ⓑ Ⓒ Ⓓ Ⓔ
33. Ⓐ Ⓑ Ⓒ Ⓓ Ⓔ
34. Ⓐ Ⓑ Ⓒ Ⓓ Ⓔ
35. Ⓐ Ⓑ Ⓒ Ⓓ Ⓔ
36. Ⓐ Ⓑ Ⓒ Ⓓ Ⓔ
37. Ⓐ Ⓑ Ⓒ Ⓓ Ⓔ
38. Ⓐ Ⓑ Ⓒ Ⓓ Ⓔ
39. Ⓐ Ⓑ Ⓒ Ⓓ Ⓔ

40. Ⓐ Ⓑ Ⓒ Ⓓ Ⓔ
41. Ⓐ Ⓑ Ⓒ Ⓓ Ⓔ
42. Ⓐ Ⓑ Ⓒ Ⓓ Ⓔ
43. Ⓐ Ⓑ Ⓒ Ⓓ Ⓔ
44. Ⓐ Ⓑ Ⓒ Ⓓ Ⓔ
45. Ⓐ Ⓑ Ⓒ Ⓓ Ⓔ
46. Ⓐ Ⓑ Ⓒ Ⓓ Ⓔ
47. Ⓐ Ⓑ Ⓒ Ⓓ Ⓔ
48. Ⓐ Ⓑ Ⓒ Ⓓ Ⓔ
49. Ⓐ Ⓑ Ⓒ Ⓓ Ⓔ
50. Ⓐ Ⓑ Ⓒ Ⓓ Ⓔ

Diagnostic Test

The Diagnostic Test is designed to help you pinpoint your weaknesses and target areas for improvement. The answer explanations that follow the test are keyed to chapters of the book. To make the best use of this diagnostic test, set aside between 1 and 2 hours so you will be able to do the whole test in one sitting. Tear out the preceding answer sheet and indicate your answers in the appropriate spaces. Complete the problems as if this were a regular testing session. Review the suggestions on pages 3–4.

When finished, check your answers against the Answer Key at the end of the test. For those that you got wrong, note the chapters containing the material that you must review. If you do not fully understand how to get a correct answer, you should review those chapters also.

Finally, fill out the Self-Evaluation Chart on page 29 in order to pinpoint the topics that gave you the most difficulty.

50 questions: 1 hour

Directions: Decide which answer choice is best. If the exact numerical value is not one of the answer choices, select the closest approximation. Fill in the circle on the answer sheet that corresponds to your choice.

Notes:
(1) You will need to use a scientific or graphing calculator to answer some of the questions.
(2) You will have to decide whether to put your calculator in degree or radian mode for some problems.
(3) All figures that accompany problems are plane figures unless otherwise stated. Figures are drawn as accurately as possible to provide useful information for solving the problem, except when it is stated in a particular problem that the figure is not drawn to scale.
(4) Unless otherwise indicated, the domain of a function is the set of all real numbers for which the functional value is also a real number.

Reference Information. The following formulas are provided for your information.

Volume of a right circular cone with radius r and height h: $V = \frac{1}{3}\pi r^2 h$

Volume of a sphere of radius r: $V = \frac{4}{3}\pi r^3$

Surface area of a sphere of radius r: $S = 4\pi r^2$

Volume of a pyramid of base area B and height h: $V = \frac{1}{3}Bh$

TIP

For the Diagnostic Test, practice exercises, and Practice Tests, an asterisk in the Answers Explained section indicates that a graphing calculator is necessary or helpful.

1. A linear function, f, has a slope of -2. $f(1) = 2$ and $f(2) = q$. Find q.

   (A) 0

   (B) $\dfrac{3}{2}$

   (C) $\dfrac{5}{2}$

   (D) 3

   (E) 4

2. A function is said to be even if $f(x) = f(-x)$. Which of the following is *not* an even function?

   (A) $y = |x|$
   (B) $y = \sec x$
   (C) $y = \log x^2$
   (D) $y = x^2 + \sin x$
   (E) $y = 3x^4 - 2x^2 + 17$

3. What is the radius of a sphere, with center at the origin, that passes through point $(2,3,4)$?

   (A) 3
   (B) 3.31
   (C) 3.32
   (D) 5.38
   (E) 5.39

4. If $f(x) = \dfrac{5x+2}{-3x+1}$, what value does $f(x)$ approach as x gets arbitrarily large?

   (A) -15

   (B) $-\dfrac{5}{3}$

   (C) -1

   (D) $\dfrac{3}{5}$

   (E) 2

5. If $f(x) = x^2 - ax$, then $f(a) =$

   (A) a
   (B) $a^2 - a$
   (C) 0
   (D) 1
   (E) $a - 1$

DIAGNOSTIC TEST

6. The average of your first three test grades is 78. What grade must you get on your fourth and final test to make your average 80?

   (A) 80
   (B) 82
   (C) 84
   (D) 86
   (E) 88

7. $\log_7 9 =$

   (A) 0.89
   (B) 0.95
   (C) 1.13
   (D) 1.21
   (E) 7.61

8. Which of the following is perpendicular to the line $y = -3x + 7$?

   (A) $y = \dfrac{1}{-3x + 7}$

   (B) $y = 7x - 3$

   (C) $y = \dfrac{1}{3}x + 5$

   (D) $y = -\dfrac{1}{3}x + 7$

   (E) $y = 3x - 7$

9. How many integers are there in the solution set of $|x - 2| \le 5$?

   (A) 0
   (B) 7
   (C) 9
   (D) 11
   (E) an infinite number

10. If $f(x) = \sqrt{x^2}$, then $f(x)$ can also be expressed as

    (A) x
    (B) $-x$
    (C) $\pm x$
    (D) $|x|$
    (E) $f(x)$ cannot be determined because x is unknown.

11. If $x + 1$ is a factor of $x^3 - 5x^2 + kx + 2$, then $k =$

    (A) -4
    (B) -2
    (C) 0
    (D) 2
    (E) 4

USE THIS SPACE FOR SCRATCH WORK

DIAGNOSTIC TEST

DIAGNOSTIC TEST 11

12. George invests $1,000 into an account that he hopes will earn 12% interest annually. How many years (rounded to the nearest year) will it take his investment to double in value?

 (A) 4
 (B) 6
 (C) 7
 (D) 8
 (E) 12

13. A linear function has an x-intercept of $\sqrt{3}$ and a y-intercept of $\sqrt{5}$. The graph of the function has a slope of

 (A) −1.29
 (B) −0.77
 (C) 0.77
 (D) 1.29
 (E) 2.24

14. If $f(x) = 2x - 1$, find the value of x that makes $f(f(x)) = 9$.

 (A) 2
 (B) 3
 (C) 4
 (D) 5
 (E) 6

15. The plane $2x + 3y - 4z = 5$ intersects the x-axis at $(a,0,0)$, the y-axis at $(0,b,0)$, and the z-axis at $(0,0,c)$. The value of $a + b + c$ is

 (A) 1
 (B) $\dfrac{35}{12}$
 (C) 5
 (D) $\dfrac{65}{12}$
 (E) 9

16. Given the set of data 1, 1, 2, 2, 2, 3, 3, 4, which one of the following statements is true?

 (A) mean ≤ median ≤ mode
 (B) median ≤ mean ≤ mode
 (C) median ≤ mode ≤ mean
 (D) mode ≤ mean ≤ median
 (E) The relationship cannot be determined because the median cannot be calculated.

17. If $\dfrac{x-3y}{x} = 7$, what is the value of $\dfrac{x}{y}$?

    (A) $-\dfrac{8}{3}$

    (B) -2

    (C) $-\dfrac{1}{2}$

    (D) $\dfrac{3}{8}$

    (E) 2

18. Find all values of x that make $\begin{vmatrix} 2 & -1 & 4 \\ 3 & 0 & 5 \\ 4 & 1 & 6 \end{vmatrix} = \begin{vmatrix} x & 4 \\ 5 & x \end{vmatrix}$.

    (A) 0
    (B) ± 1.43
    (C) ± 3
    (D) ± 4.47
    (E) 5.34

19. Suppose $f(x) = \dfrac{1}{2}x^2 - 8$ for $-4 \le x \le 4$. Then the maximum value of the graph of $|f(x)|$ is

    (A) -8
    (B) 0
    (C) 2
    (D) 4
    (E) 8

20. If $\tan \theta = \dfrac{2}{3}$, then $\sin \theta =$

    (A) ± 0.55
    (B) ± 0.4
    (C) 0.55
    (D) 0.83
    (E) 0.89

21. Suppose the graph of $f(x) = -x^3$ is translated 4 units right and 2 units down, resulting in the graph of a new function g. What is the value of $g(-2)$?

    (A) -218
    (B) -10
    (C) 6
    (D) 214
    (E) 218

22. $i^{2014} =$

(A) i^{13}
(B) i^{203}
(C) i^{726}
(D) i^{1993}
(E) i^{2100}

23. The statistics below provide a summary of IQ scores of 100 children.

Mean: 100
Median: 102
Standard Deviation: 10
First Quartile: 84
Third Quartile: 110

About 50 of the children in this sample have IQ scores that are

(A) less than 84
(B) less than 110
(C) between 84 and 110
(D) between 64 and 130
(E) more than 100

24. If $f(x) = \sec x$, then

(A) $f(x) = f(-x)$

(B) $f\left(\dfrac{1}{x}\right) = -f(x)$

(C) $f(-x) = -f(x)$

(D) $f(x) = f\left(\dfrac{1}{x}\right)$

(E) $f(x) = \dfrac{1}{f(x)}$

25. The polar coordinates of a point P are $(2, 240°)$. The Cartesian (rectangular) coordinates of P are

(A) $\left(-1, -\sqrt{3}\right)$

(B) $\left(-1, \sqrt{3}\right)$

(C) $\left(-\sqrt{3}, -1\right)$

(D) $\left(-\sqrt{3}, 1\right)$

(E) $\left(1, -\sqrt{3}\right)$

26. The height of a cone is equal to the radius of its base. The radius of a sphere is equal to the radius of the base of the cone. The ratio of the volume of the *cone* to the volume of the *sphere* is

(A) $\dfrac{1}{12}$

(B) $\dfrac{1}{4}$

(C) $\dfrac{1}{3}$

(D) $\dfrac{1}{1}$

(E) $\dfrac{4}{3}$

27. In how many distinguishable ways can the seven letters in the word MINIMUM be arranged, if all the letters are used each time?

(A) 7
(B) 42
(C) 420
(D) 840
(E) 5,040

28. Which of the following lines are asymptotes of the graph of $y = \dfrac{x}{x+1}$?

  I. $x = 1$
 II. $x = -1$
III. $y = 1$

(A) I only
(B) II only
(C) III only
(D) I and II
(E) II and III

29. What is the probability of getting at least three heads when flipping four coins?

(A) $\dfrac{3}{16}$

(B) $\dfrac{1}{4}$

(C) $\dfrac{5}{16}$

(D) $\dfrac{7}{16}$

(E) $\dfrac{3}{4}$

30. How many four-digit personal identification numbers are possible if no digit can be used twice and no number can begin with 0?

(A) 210
(B) 4,536
(C) 6,561
(D) 10,000
(E) 362,880

31. In the figure above, S is the set of all points in the shaded region. Which of the following represents the set consisting of all points (2x,y), where (x,y) is a point in S?

(A)

(B)

(C)

(D)

(E)

32. If a square prism is inscribed in a right circular cylinder of radius 3 and height 6 as shown above, the volume inside the cylinder but outside the prism is

(A) 2.14
(B) 3.14
(C) 61.6
(D) 115.6
(E) 169.6

33. What is the length of the major axis of the ellipse whose equation is $10x^2 + 20y^2 = 200$?

(A) 3.16
(B) 4.47
(C) 6.32
(D) 8.94
(E) 14.14

34. The fifth term of an arithmetic sequence is 26, and the eighth term is 41. What is the first term?

(A) 3
(B) 4
(C) 5
(D) 6
(E) 7

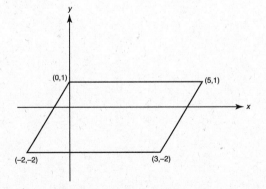

35. What is the measure of one of the larger angles of the parallelogram above that has vertices at $(-2,-2)$, $(0,1)$, $(5,1)$, and $(3,-2)$?

(A) 117.2°
(B) 123.7°
(C) 124.9°
(D) 125.3°
(E) 131.0°

36. If $\cos x = -0.25$, then $\cos(\pi - x) =$

    (A) -0.25
    (B) -0.75
    (C) 0
    (D) 0.75
    (E) 0.25

37. The parametric equations of a line are $x = 3 - t$ and $y = 1 + t$. The slope of this line is

    (A) -3
    (B) -1
    (C) $-\dfrac{1}{3}$
    (D) 1
    (E) 3

38. If $f(x) = 2x^2 - 4$ and $g(x) = 2^x$, the value of $g(f(1))$ is

    (A) -4
    (B) 0
    (C) $\dfrac{1}{4}$
    (D) 1
    (E) 4

39. If $f(x) = 3\sqrt{5x}$, what is the value of $f^{-1}(15)$?

    (A) 0.65
    (B) 0.90
    (C) 5.00
    (D) 7.5
    (E) 25.98

40. If $2 \sin^2 x - 3 = 3 \cos x$ and $90° < x < 270°$, the number of values that satisfy the equation is

    (A) 0
    (B) 1
    (C) 2
    (D) 3
    (E) 4

41. Thirty percent of the 20 people in the Math Club have blonde hair. If 3 people are selected at random from the club, what is the probability that none has blonde hair?

    (A) 0.1
    (B) 0.25
    (C) 0.32
    (D) 0.40
    (E) 0.50

DIAGNOSTIC TEST

42. Which of the following could be the equation of one cycle of the graph in the figure above?

I. $y = \sin 4x$

II. $y = \cos\left(4x - \dfrac{\pi}{2}\right)$

III. $y = -\sin(4x + \pi)$

(A) only I
(B) only I and II
(C) only II and III
(D) only II
(E) I, II, and III

43. Observers at locations due north and due south of a rocket launchpad sight a rocket at a height of 10 kilometers. Assume that the curvature of Earth is negligible and that the rocket's trajectory at that time is perpendicular to the ground. How far apart are the two observers if their angles of elevation to the rocket are 80.5° and 68.0°?

(A) 0.85 km
(B) 4.27 km
(C) 5.71 km
(D) 20.92 km
(E) 84.50 km

44. The vertex angle of an isosceles triangle is 35°. The length of the base is 10 centimeters. How many centimeters are in the perimeter?

(A) 16.6
(B) 17.4
(C) 20.2
(D) 43.3
(E) 44.9

45. If the graph below represents the function $y = f(x)$, which of the following could represent the equation of the inverse of f?

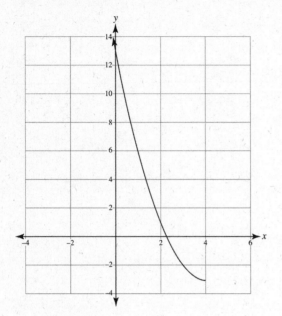

(A) $x = y^2 - 8y - 1$
(B) $x = y^2 + 11$
(C) $x = (y - 4)^2 - 3$
(D) $x = (y + 4)^2 - 3$
(E) $x = (y + 4)^2 + 3$

46. If $k > 4$ is a constant, how would you translate the graph of $y = x^2$ to get the graph of $y = x^2 + 4x + k$?

(A) left 2 units and up k units
(B) right 2 units and up $(k - 4)$ units
(C) left 2 units and up $(k - 4)$ units
(D) right 2 units and down $(k - 4)$ units
(E) left 2 units and down $(k - 4)$ units

47. If $f(x) = \log_b x$ and $f(2) = 0.231$, the value of b is

(A) 0.3
(B) 1.3
(C) 13.2
(D) 20.1
(E) 32.5

48. If $a_{n+1} = a_{n-1} + 2a_n$ for $n = 2, 3, 4, \ldots$, and $a_1 = 1$ and $a_2 = 1$, then $a_5 =$

(A) 7
(B) 11
(C) 17
(D) 21
(E) 41

USE THIS SPACE FOR SCRATCH WORK

49. Suppose $\cos \theta = u$ in $0 < \theta < \frac{\pi}{2}$. Then $\tan \theta =$

(A) 1

(B) $\dfrac{1}{\sqrt{1-u^2}}$

(C) $\dfrac{u}{\sqrt{1-u^2}}$

(D) $\sqrt{1-u^2}$

(E) $\dfrac{\sqrt{1-u^2}}{u}$

50. A certain component of an electronic device has a probability of 0.1 of failing. If there are 6 such components in a circuit, what is the probability that at least one fails?

(A) 0.60
(B) 0.47
(C) 0.167
(D) 0.000006
(E) 0.000001

STOP

If there is still time remaining, you may review your answers.

ANSWER KEY
Diagnostic Test

1. A	14. B	27. C	40. D
2. D	15. B	28. E	41. C
3. E	16. C	29. C	42. E
4. B	17. C	30. B	43. C
5. C	18. D	31. C	44. D
6. D	19. E	32. C	45. C
7. C	20. A	33. D	46. C
8. C	21. D	34. D	47. D
9. D	22. C	35. B	48. C
10. D	23. C	36. E	49. E
11. A	24. A	37. B	50. B
12. B	25. A	38. C	
13. A	26. B	39. C	

ANSWERS EXPLAINED

The following explanations are keyed to the review portions of this book. The number in brackets after each explanation indicates the appropriate chapter in the Review of Major Topics (Part 2).

An asterisk appears next to those solutions for which a graphing calculator is necessary or helpful.

1. **(A)** $f(1) = 2$ means that the line goes through point $(1,2)$. $f(2) = q$ means that the line goes through point $(2,q)$. Slope $= \dfrac{\Delta y}{\Delta x} = \dfrac{q-2}{2-1} = -2$ implies $-2 = \dfrac{q-2}{1}$, so $q = 0$. [2]

* 2. **(D)** Even functions are symmetric about the y-axis. Graph each answer choice to see that Choice D is not symmetric about the y-axis.

An alternative solution is to use the fact that $\sin x \neq \sin(-x)$, from which you deduce the correct answer choice. [1]

Note: box on right

> **IMPORTANT REMINDER**
>
> Properties of even and odd functions:
>
> Even + even and even × even are always an even function.
>
> Odd + odd is always an odd function, but odd × odd is an even function.
>
> Odd × even is always an odd function.

* 3. **(E)** Since the radius of a sphere is the distance between the center, $(0,0,0)$, and a point on the surface, $(2,3,4)$, use the distance formula in three dimensions to get:

$$\sqrt{(2-0)^2 + (3-0)^2 + (4-0)^2} = \sqrt{29}$$

Use your calculator to find $\sqrt{29} \approx 5.39$. [15]

4. **(B)** The polynomials in the numerator and the denominator of this rational function have the same degree. Therefore, as x gets arbitrarily large, $f(x)$ approaches the ratio of the leading coefficients, $-\dfrac{5}{3}$. [9]

5. **(C)** $f(a)$ means to replace x in the formula with an a. Therefore, $f(a) = a^2 - a \cdot a = 0$. [1]

* 6. **(D)** Since the average of your first three test grades is 78, each test grade could have been a 78. If x represents your final test grade, the average of the four test grades is $\dfrac{78 + 78 + 78 + x}{4}$, which is to be equal to 80. Therefore, $\dfrac{234 + x}{4} = 80$.

$234 + x = 320$. So $x = 86$. [21]

7. **(C)** Use the change-of-base theorem and your calculator to get:

$$\log_7 9 = \frac{\log_{10} 9}{\log_{10} 7} \approx \frac{0.9542}{0.8451} \approx 1.13$$

In some of the newer TI-84$^+$ calculators, you may enter the base, resulting in a single calculation. [8]

8. **(C)** The slope of the given line is -3. Therefore, the slope of a perpendicular line is the negative reciprocal, or $\dfrac{1}{3}$. [2]

✳ 9. **(D)** Plot the graph of $y = \text{abs}(x-2) - 5$ in the standard window that includes both x-intercepts. You can count 11 integers between -3 and 7 if you include both endpoints.

The inequality $|x - 2| \leq 5$ means that x is less than or equal to 5 units away from 2. Therefore, $-3 \leq x \leq 7$, and there are 11 integers in this interval. [11]

10. **(D)** $\sqrt{x^2}$ indicates the need for the *positive* square root of x^2. Therefore, $\sqrt{x^2} = x$ if $x \geq 0$ and $\sqrt{x^2} = -x$ if $x < 0$. This is just the definition of absolute value, and so $\sqrt{x^2} = |x|$ is the only answer for all values of x. [11]

11. **(A)** If $x + 1$ is a factor of $x^3 - 5x^2 + kx + 2$, then substituting -1 for x should leave a remainder of 0: $-1 - 5 - k + 2 = 0$. Therefore, $k = -4$. [4]

✳ 12. **(B)** George's investment doubles to \$2,000. Therefore, $2{,}000 = 1{,}000(1 + 0.12)^t$, or $2 = 1.12^t$. Take the log of both sides to get $t = \log_{1.12} 2 = \dfrac{\log 2}{\log 1.12} \approx 6.12 \approx 6$ years. [8]

✳ 13. **(A)** $y = mx + b$. Use the x-intercept to get $0 = \sqrt{3}m + b$, and the y-intercept to get $\sqrt{5} = 0 \cdot m + b$. Therefore, $0 = \sqrt{3}m + \sqrt{5}$ and $m = -\dfrac{\sqrt{5}}{\sqrt{3}} \approx -1.29$. [2]

14. **(B)** $f(f(x)) = 2(2x - 1) - 1 = 4x - 3$. Solve $4x - 3 = 9$ to get $x = 3$. [1]

15. **(B)** Substituting the points into the equation gives $a = \dfrac{5}{2}$, $b = \dfrac{5}{3}$, and $c = -\dfrac{5}{4}$, so $a + b + c = \dfrac{35}{12}$. [15]

16. **(C)** Mode $= 2$, median $= \dfrac{2+2}{2} = 2$, mean $= \dfrac{2+6+6+4}{8} = \dfrac{18}{8} = 2.25$. Thus, median $\leq$ mode $\leq$ mean. [21]

17. **(C)** Multiply $\dfrac{x - 3y}{x} = 7$ by x to get $x - 3y = 7x$. Subtract x from both sides to get $-3y = 6x$. Divide by $6y$ so that $\dfrac{x}{y}$ will be on one side of the equal sign. This gives $\dfrac{x}{y} = -\dfrac{3}{6} = -\dfrac{1}{2}$. [1]

✳ 18. **(D)** Enter the 3 by 3 matrix into the graphing calculator and evaluate its determinant as 0. The 2 by 2 matrix on the right side of the equation has the determinant $x^2 - 20$. Solve this for x to get ± 4.47. [18]

✳ 19. **(E)** Plot the graph of $y = \text{abs}((1/2)x^2 - 8)$ in a $[-4,4]$ by $[-10,10]$ window and observe that the maximum value is 8 (at $x = 0$).

An alternative solution is to recognize that the graph of f is a parabola that is symmetric about the y-axis and opens up. The minimum value $y = -8$ occurs when $x = 0$, so 8 is the maximum value of $|f(x)|$. [3]

* **20.** **(A)** Tan is positive in the first and third quadrants, so θ is a first or third quadrant angle. The sine of a first quadrant angle is positive, but the sine of a third quadrant angle is negative. Enter $\sin(\tan^{-1}(2/3)) \approx 0.55$. The correct answer choice is therefore ± 0.55. [7]

* **21.** **(D)** The function is $g(x) = f(x - 4) - 2 = -(x - 4)^3 - 2$. Enter $g(x)$ as Y_1 in the graphing calculator and find $g(-2) = 214$. [12]

22. **(C)** Powers of i repeat in fours. Therefore, two powers of i are equal if their remainders are equal upon division by four. When 2014 is divided by 4, the remainder is 2. When 726 is divided by 4, the remainder is also 2. [17]

23. **(C)** Fifty children is half of 100 children, and half of the data points lie between the first and third quartiles. [21]

* **24.** **(A)** Plot the graph of $y = \dfrac{1}{\cos x}$ and observe that it is symmetric about the y-axis. Hence $f(x) = \sec x$ is an even function and $f(x) = f(-x)$. An alternative solution is to recall that $\cos x$ is even, so its reciprocal is also even. [1]

25. **(A)** From the figure below you can see that the Cartesian (rectangular) coordinates of P are $\left(-1, -\sqrt{3}\right)$. [14]

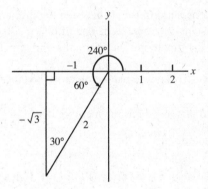

26. **(B)** Since $r = h$ in the cone,

$$\frac{\text{Volume of cone}}{\text{Volume of sphere}} = \frac{\dfrac{1}{3}\pi r^2 h}{\dfrac{4}{3}\pi r^3} = \frac{\dfrac{1}{3}\pi r^3}{\dfrac{4}{3}\pi r^3} = \frac{1}{4}. \ [15]$$

* **27.** **(C)** The word MINIMUM contains 7 letters, which can be permuted 7! ways. The 3 Ms can be permuted 3! ways, and the 2 Is can be permuted in 2! ways, so only $\dfrac{1}{3!2!} = \dfrac{1}{12}$ permutations look different from each other. Therefore, there are $\dfrac{7!}{12} = 420$ distinguishable ways the letters can be arranged. [16]

28. **(E)** Vertical asymptotes occur when the denominator is zero, so $x = -1$ is the only vertical asymptote. Since $\dfrac{x}{x + 1}$ gets closer to 1 as x gets larger, the horizontal asymptote is $y = 1$. [9]

* **29.** **(C)** Since each of the 4 flips has 2 possible outcomes (heads or tails), there are $2^4 = 16$ outcomes in the sample space. At least 3 heads means 3 or 4 heads.

$\begin{pmatrix} 4 \\ 3 \end{pmatrix} = 4$ ways to get 3 heads. $\begin{pmatrix} 4 \\ 4 \end{pmatrix} = 1$ way to get 4 heads and $\frac{4+1}{16} = \frac{5}{16}$. [22]

* **30.** **(B)** There are 9 digits that could go in the first position. For each of those, there are 9 that could go in the second position; 8 that could go in the third position; and 7 that could go in the fourth position. Therefore, there are $9 \times 9 \times 8 \times 7 = 4{,}536$ four-digit numbers. [16]

31. **(C)** Since the y-values remain the same but the x-values are doubled, the circle is stretched along the x-axis. [12]

* **32.** **(C)** Volume of cylinder $= \pi r^2 h = \pi \cdot 9 \cdot 6 = 54\pi$. Volume of square prism $= Bh$, where B is the area of the square base, which is $3\sqrt{2}$ on a side. Thus, $Bh = (3\sqrt{2})^2 \cdot 6 = 108$. Therefore, the desired volume is $54\pi - 108$, which (using your calculator) is approximately 61.6. [15]

33. **(D)** Divide both sides of the equation by 200 to write the equation in standard form $\frac{x^2}{20} + \frac{y^2}{10} = 1$. In this equation, $a = \sqrt{20}$. The length of the major axis is $2\sqrt{20} \approx 8.94$. [13]

34. **(D)** There are three constant differences between the fifth and eighth terms. Since $41 - 26 = 15$, the constant difference is 5. The fifth term, 26, is four constant differences (20) more than the first term. Therefore, the first term is $26 - 20 = 6$. [19]

* **35.** **(B)** The figure on page 17 shows a parallelogram. The tangent of the acute angle with vertex $(-2,-2)$ is $\frac{3}{2}$, so that angle has the measure $\tan^{-1}\left(\frac{3}{2}\right) = 56.3°$. The larger angle is the supplement, or 123.7°. [6]

36. **(E)** If $\cos x$ is negative, then x must be in QII or QIII. If x is in QII, then $-x$ is in QIII, so $\pi - x$ is QI. If x is in QIII, then $-x$ is in QII, so $\pi - x$ is in QIV. Therefore, $\cos(\pi - x) = 0.25$. You could also key in $\cos^{-1}(-0.25)$ and then key in $\cos(\pi - \text{Ans})$ to get 0.25. [7]

* **37.** **(B)** Solve $x = 3 - t$ for t to get $t = 3 - x$, and substitute this in $y = 1 + t$ to get $y = 1 + 3 - x = -x + 4$. The slope of this line is -1. [10]

38. **(C)** $f(1) = 2 - 4 = -2$, and $g(-2) = 2^{-2} = \frac{1}{4}$. [1]

39. **(C)** $f^{-1}(15)$ is the value of x that makes $3\sqrt{5x}$ equal to 15. Set $3\sqrt{5x} = 15$, and divide both sides by 3 to get $\sqrt{5x} = 5$. Therefore, $5x = 25$, and $x = 5$. [1]

∗ 40. **(D)** With your calculator in degree mode, plot the graphs of $2(\sin(x))^2 - 3$ and $y = 3\cos x$ in a [90,270] by [–3,0] window and observe that the graphs intersect in 3 places.

An alternative solution is to distribute and transform the equation to read as:

$$2\cos^2 x + 3\cos x + 1 = 0$$

This factors to $(2\cos x + 1)(\cos x + 1) = 0$, so $\cos x = -\dfrac{1}{2}$ or $\cos x = -1$.

For $90° < x < 270°$, there are three solutions: $x = 120°, 240°, 180°$. [7]

∗ 41. **(C)** Thirty percent of 20 is 6. Therefore, 14 people in the Math Club don't have blonde hair. There are $_{14}C_3$ ways of choosing 3 of these 14 people and $_{20}C_3$ ways of choosing 3 of 20 people. The ratio of these two numbers is $\dfrac{364}{1140} = 0.32$, the probability of choosing 3 people who don't have blonde hair. [22]

∗ 42. **(E)** Plot the graphs of all three functions in a $\left[-\dfrac{\pi}{4}, \dfrac{\pi}{4}\right]$ by [–2,2] window and observe that they coincide.

An alternative solution is to deduce facts about the graphs from the equations. All three equations indicate graphs that have period $\dfrac{\pi}{2}$. The graph of equation I is a normal sine curve. The graph of equation II is a cosine curve with a phase shift right of $\dfrac{\pi}{8}$, one-fourth of the period. Therefore, it fits a normal sine curve. The graph of equation III is a sine curve that has a phase shift left of $\dfrac{\pi}{4}$, one-half the period, and reflected through the x-axis. This also fits a normal sine curve. [7]

∗ 43. **(C)** The given information is illustrated in the figure below.

Points A and B represent the two observers. Point C is the rocket, and point D is the base of the altitude from the rocket to the ground. We know that $\tan 80.5° = \dfrac{10}{AD}$ and $\tan 68.0° = \dfrac{10}{DB}$. Therefore, $AB = AD + DB = \dfrac{10}{\tan 80.5°} + \dfrac{10}{\tan 68.0°} \approx 5.71$ km. [6]

∗ 44. **(D)** Drop the altitude from the vertex to the base. The altitude bisects both the vertex angle and the base, cutting the triangle into two congruent right triangles.

Since $\sin 17.5° = \dfrac{5}{\text{leg}}$, leg $= \dfrac{5}{\sin 17.5°} \approx \dfrac{5}{0.3007} \approx 16.628$ cm and the

perimeter $= 2(16.628) + 10 = 43.3$ cm. [6]

45. **(C)** The given graph looks like the left half of a parabola with vertex $(4,-3)$ that opens up. The equation of such a parabola is $y = (x - 4)^2 - 3$. The vertex of the inverse is $(-3,4)$, so its equation is $x = (y - 4)^2 - 3$. [3]

46. **(C)** Complete the square of $x^2 + 4x + k$ by adding and subtracting 4 to get the translated function of $y = (x + 2)^2 + (k - 4)$. Translate $y = x^2$ left 2 units and up $(k - 4)$ units. [12]

∗ 47. **(D)** $f(2) = \log_b 2 = 0.231$. Therefore, $b^{0.231} = 2$. Taking $\log_2$ of both sides yields $0.231 \log_2 b = 1$, so $\log_2 b = \dfrac{1}{0.231}$. Therefore, $b = 2^{1/0.231} = 20.1$. [8]

48. **(C)** $a_3 = a_1 + 2a_2 = 1 + 2 = 3$; $a_4 = a_2 + 2a_3 = 1 + 6 = 7$; $a_5 = a_3 + 2a_4 = 3 + 14 = 17$. [19]

49. **(E)** Since $0 < \theta < \dfrac{\pi}{2}$, the figure below shows θ in Quadrant I with $\cos\theta = u$.

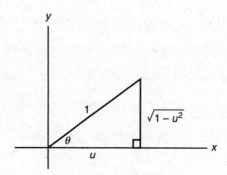

From the figure, $\tan\theta = \dfrac{\sqrt{1 - u^2}}{u}$. [7]

∗ 50. **(B)** The probability that at least one component fails is 1 minus the probability that all succeed. Since the probability of one component succeeding is 1 minus 0.1, or 0.9, the probability that all succeed is $(0.9)^6 = 0.53$, and $1 - 0.53 = 0.47$. [22]

Subject Area	Questions and Review Section							Right	Number Wrong	Omitted
Algebra and Functions	1	2	4	5	7	8				
(22 questions)	2	1	9	1	8	2		____	____	____
	9	10	11	12	13	14	17			
	11	11	4	8	2	1	1	____	____	____
	19	21	24	28	37	38	39			
	3	12	1	9	10	1	1	____	____	____
	45	47								
	3	8						____	____	____
Trigonometry	20	35	36	40	42	43				
(8 questions)	7	6	7	7	7	6		____	____	____
	44	49								
	6	7						____	____	____
Coordinate and Three-Dimensional Geometry	3	15	25	26	31	32				
	15	15	14	15	12	15		____	____	____
(8 questions)										
	33	46								
	13	12						____	____	____
Number and Operations	18	22	27	30	34	48				
(6 questions)	18	17	16	16	19	19		____	____	____
Data Analysis, Statistics, and Probability	6	16	23	29	41	50				
	21	21	21	22	22	22		____	____	____
(6 questions)										
TOTALS								____	____	____

Evaluate Your Performance on the Diagnostic Test	
Rating	**Number Right**
Excellent	41–50
Very good	33–40
Above average	25–32
Average	15–24
Below average	Below 15

CALCULATING YOUR SCORE

Raw score R = number right $- \frac{1}{4}$ (number wrong), rounded = _____

Approximate scaled score $S = 800 - 10(44 - R) =$ _____

If $R \geq 44$, $S = 800$.

PART 2
Review of Major Topics

This part reviews the mathematical concepts and techniques for the topics covered in the Level 2 Mathematics Subject Test. A sound understanding of these concepts certainly will improve your score. The techniques discussed may help you save time solving some of the problems without a calculator at all. For problems requiring computational power, techniques are described that will help you use your calculator in the most efficient manner.

Your classroom experience will guide your decisions about how best to use a graphing calculator. If you have been through a secondary mathematics program that attached equal importance to graphical, tabular, and algebraic presentations, then you probably will rely on your graphing calculator as your primary tool to help you find solutions. However, if you went through a more traditional mathematics program, where algebra and algebraic techniques were stressed, it may be more natural for you to use a graphing calculator only after considering other approaches.

Functions

→ **DEFINITIONS**
→ **COMBINING FUNCTIONS**
→ **PIECEWISE FUNCTIONS**
→ **INVERSES**
→ **ODD AND EVEN FUNCTIONS**

A **function** is a process that changes a set of *input* numbers into a set of *output* numbers. A particular input number will always produce the same output number. However, a particular output number could be the result of applying the function to different input numbers. Functions are usually specified by equations such as $y = \sqrt{2x - 1}$. In this specific equation, x represents an input number, while y represents the (unique) corresponding output number. Functions can also have names: in the example, we could name the function f. Then the process could be described as $f(x) = \sqrt{2x - 1}$, whereby f takes the input number x, multiplies it by 2, subtracts 1, and takes the square root to produce the output $y = f(x)$.

Taken as a group, the input numbers are called the **domain** of the function, while the output numbers are called the **range**. Unless otherwise specified, the domain of a function is all real numbers for which the equation produces outputs that are real numbers. In the example above, the domain is the set $x \geq \dfrac{1}{2}$ since $2x - 1$ cannot be negative if y is to be a real number. In this case, the range is the set of all non-negative numbers.

The domain of a function can also be established as part of the definition of a function. For example, even though the domain of the example function f is $x \geq \dfrac{1}{2}$, one could specify the domain $x > 5$. Unless a domain is explicitly stated, the domain is assumed to be all real values that produce real numbers as outputs.

A function with a small finite domain can be described by a set of ordered pairs instead of an equation. The first number in the pair is from the domain and the second is the corresponding range value. Consider, for example, a function f consisting of the pairs $(0, 2)$, $(1, 1)$, $(2, 3)$, and $(3, 8)$. In this example, the "process" is not systematic: it simply changes 0 to 2 ($f(0) = 2$); 1 to 1 ($f(1) = 1$); 2 to 3 ($f(2) = 3$); and 3 to 8 ($f(3) = 8$). The domain of this function consists of 0, 1, 2, and 3, while the range consists of 2, 1, 3, and 8. Functions like this are typically used to illustrate certain properties of functions and are discussed later.

A function is actually a special type of **relation**. A relation describes the association between two variables. An equation such as $x^2 + y^2 = 4$ is one way of defining a relation. All ordered pairs (x, y) that satisfy the equation are in the relation. In this case, these pairs form the circle of radius 2 centered at the origin. Note that in a relation, unlike a function, a particular input number might be paired with more than one output number.

Circles are examples of relations that are not functions because some x-values (0 in the example) have two y-values associated with it (2 and –2), which violates the uniqueness of the output for a given input. Other than circles, relations that are not functions include ellipses, hyperbolas, and parabolas that open right or left, instead of up or down—in other words, the conic sections discussed in Chapter 13.

Like functions, relations can also be defined using specific ordered pairs. The set $R = \{(1, 1), (1, 2), (2, 1), (2, 2)\}$ is an example of a relation that is not a function because the x-value 1 has two y-values associated with it (1 and 2).

A graph is the graph of a function if it passes the Vertical Line Test. Since a vertical line represents an x-value, a graph passes this test if no vertical line intersects the graph in more than one point.

TIP

Typically, a value of *x* that must be excluded from the domain of a function is one that makes the denominator zero or makes the value of an expression under a square root less than zero.

Given two functions, f and g, six new functions can be defined:

Sum function	$(f + g)(x) = f(x) + g(x)$
Difference function	$(f - g)(x) = f(x) - g(x)$
Product function	$(f \cdot g)(x) = f(x) \cdot g(x)$
Quotient function	$\left(\dfrac{f}{g}\right)(x) = \dfrac{f(x)}{g(x)}, g(x) \neq 0$
Composition of functions	$(f \circ g)(x) = f(g(x))$
	$(g \circ f)(x) = g(f(x))$

➡ **Examples**_____

If $f(x) = 3x - 2$ and $g(x) = x^2 - 4$, write an expression for each of the following functions:

(A) $(f + g)(x)$

$$(f + g)(x) = f(x) + g(x)$$
$$= (3x - 2) + (x^2 - 4) = x^2 + 3x - 6$$

(B) $(f - g)(x)$

$$(f - g)(x) = f(x) - g(x)$$
$$= (3x - 2) - (x^2 - 4) = -x^2 + 3x + 2$$

(C) $(f \cdot g)(x)$

$$(f \cdot g)(x) = f(x) \cdot g(x)$$
$$= (3x - 2)(x^2 - 4)$$
$$= 3x^3 - 2x^2 - 12x + 8$$

(D) $\left(\dfrac{f}{g}\right)(x)$

$$\left(\dfrac{f}{g}\right)(x) = \dfrac{f(x)}{g(x)} = \dfrac{3x - 2}{x^2 - 4} \text{ and } x \neq \pm 2$$

(E) $(f \circ g)(x)$

$$(f \circ g)(x) = f(g(x)) = 3(g(x)) - 2$$
$$= 3(x^2 - 4) - 2 = 3x^2 - 14$$

(F) $(g \circ f)(x)$

$$(g \circ f)(x) = g(f(x)) = (f(x))^2 - 4$$
$$= (3x - 2)^2 - 4 = 9x^2 - 12x$$

The inverse of a function f, denoted by f^{-1}, is a relation that has the property that $(f^{-1} \circ f)(x) = (f \circ f^{-1})(x) = x$ for all x in the domain of f. The next two examples show how to find the inverse of a function and verify that it is the inverse.

➡ **Examples**_____

(G) Given $f(x) = 3x + 2$. Find $f^{-1}(x)$.

First write y in place of $f(x)$ and interchange x and y. Then solve for y and call the result $g(x)$:

$$x = 3y + 2 \text{ so } y = g(x) = \dfrac{x - 2}{3}$$

g is the candidate for f^{-1}.

TIP

$(f \circ g)(x)$ and $(g \circ f)(x)$ need *not be* the same!

(H) $f(x) = 3x + 2$. Is $g(x) = \dfrac{x-2}{3}$ the inverse of f?

$$(g \circ f)(x) = g\big(f(x)\big) = g(3x+2) = \frac{3x+2-2}{3} = x$$

$$(f \circ g)(x) = f\big(g(x)\big) = 3\left(\frac{x-2}{3}\right) + 2 = x$$

Therefore, $g = f^{-1}$.

The inverse of a function is not necessarily a function.

➥ Examples

(I) $f(x) = x^2$. Find f^{-1}.

Write y in place of $f(x)$ and interchange x and y. Then solve for y and call the result $g(x)$:

$$x = y^2 \;\text{ so } y = g(x) = \pm\sqrt{x}$$

Since a given input x produces two output values, $\sqrt{x}$ and $-\sqrt{x}$, g is not a function.

(J) Find the inverse of $f = \{(1,2),(2,3),(3,2)\}$.

Interchange x and y to get $f^{-1} = \{(2,1),(3,2),(2,3)\}$. Since the input 2 produces two outputs (1 and 3), f^{-1} is not a function.

If the inverse of a function is not a function, as in Example I, it can be made into a function by limiting the domain of the original function. In the example, the domain of f could have been limited to $x \geq 0$ or $x \leq 0$. Then the inverse would have been $\sqrt{x}$ or $-\sqrt{x}$, respectively.

If a point with coordinates (a, b) belongs to a function, then the point with coordinates (b, a) belongs to its inverse. Therefore, the graphs of inverses are reflections about the line $y = x$. This is illustrated for $f(x) = x^2$ in the figure below. The inverse in this figure is graphed as a dashed curve. You can see that the inverse is not a function because it doesn't pass the Vertical Line Test. Thus a function does not have an inverse if there is a horizontal line that hits its graph in more than one point.

Since the domains and ranges of inverses are switched, the range of a function can be found by finding the domain of its inverse.

➥ Example

(K) Find the range of the function $f(x) = \dfrac{1}{x+4} - 2$.

Using the method described in the previous examples, find the inverse to be $f^{-1}(x) = \dfrac{1}{x+2} - 4$.

Since $x = -2$ is not in the domain of f^{-1}, the value -2 is not in the range of f, so the range of f^{-1} is all real numbers except -2.

A function f is called **even** if $f(-x) = f(x)$ for all x in its domain. For even functions, outputs are equal for opposite inputs. In fact, the designation "even" comes from power functions of the form $f(x) = x^n$, where n is an even integer. The graphs of even functions are symmetric about the y-axis.

A function f is called **odd** if $f(-x) = -f(x)$ for all x in its domain. For odd functions, outputs are opposite for opposite inputs. Again, the designation "odd" comes from power functions $f(x) = x^n$, where n is an odd integer. The graphs of odd functions are symmetric about the origin.

Functions need not be even or odd. A function such as $f(x) = x^2 + x + 1$ is neither even nor odd. At the other extreme, $f(x) = 0$ is both even and odd.

Even and odd relations are similarly defined. A relation is called even if $(-x, y)$ is in the relation whenever (x, y) is, and it is called odd if $(-x, -y)$ is in the relation whenever (x, y) is. The unit circle $x^2 + y^2 = 1$ is an example of a relation that is both even and odd.

The sum of even functions or relations is even, and the sum of odd functions is odd. The product of two even or odd functions is even, while the product of an even function and an odd function is odd.

➡ **Examples** _____

(L) $f(x) = x^2$ and $f(-x) = (-x)^2 = x^2$. **Therefore, f is even.**

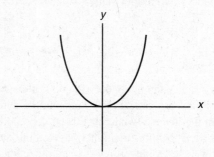

(M) $f(x) = |x|$ and $f(-x) = |-x| = |-1 \cdot x| = |-1| \cdot |x| = |x|$. **Therefore, f is even.**

(N) $f(x) = \cos x$ **is an even function because** $\cos(-x) = \cos x$.

(O) $f(x) = x^3$ and $f(-x) = (-x)^3 = -x^3$.

Therefore, $f(-x) = x^3 = -f(x)$, so f is odd.

(P) $f(x) = \dfrac{1}{x}$ and $f(-x) = \dfrac{1}{-x}$.

Therefore, $f(-x) = -\dfrac{1}{x} = -f(x)$, so f is odd.

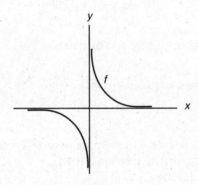

(Q) $f(x) = \sin x$ is odd because $\sin(-x) = -\sin x$.

TIP

Relations can be odd, even, or neither. They can also be both odd and even!

A **piecewise function** is a function that is defined by different equations on different parts of its domain. Piecewise functions provide useful illustrations of calculus concepts, but they also occur in real-life situations.

➡ Examples

(R) The function f is defined by $f(x) = \begin{cases} x+3 & \text{if } x < 1 \\ x^2 & \text{if } x \geq 1 \end{cases}$. Find $f(-5)$, $f(1)$, and $f(3)$.

$$f(-5) = -5 + 3 = -2$$
$$f(1) = 1^2 = 1$$
$$f(3) = 3^2 = 9$$

(S) A store that makes copies charges 25 cents per copy per page if 50 or fewer copies of each page are made; 20 cents per copy if between 51 and 200 copies of each page are made; and 15 cents per copy if more than 200 copies of each page are made. How would this be written as a piecewise function?

$$f(x) = \begin{cases} 0.25x & \text{if } x \leq 50 \\ 0.20x & \text{if } 51 \leq x \leq 200 \\ 0.15x & \text{if } x > 200 \end{cases}$$

In this example, x would have to be a counting number.

(T) Find the range of the piecewise function $f(x) = \begin{cases} 2^x & \text{if } x \leq 3 \\ x^2+1 & \text{if } x > 3 \end{cases}$ and write it in interval notation.

The range of 2^x for $x \leq 3$ is $0 < y \leq 8$, and the range of $x^2 + 1$ for $x > 3$ is $10 < y < \infty$. The union of these two ranges in interval notation is $(0,8] \cup [10,\infty)$. There is a "gap" between 8 and 10 in the range of this function.

Piecewise functions are discussed at greater length in Chapter 11.

1. If {(3,2),(4,2),(3,1),(7,1),(2,3)} is to be a function, which one of the following must be removed from the set?

    (A) (3,2)
    (B) (4,2)
    (C) (2,3)
    (D) (7,1)
    (E) none of the above

2. For $f(x) = 3x^2 + 4$, $g(x) = 2$, and $h = \{(1,1),(2,1),(3,2)\}$,

    (A) f is the only function
    (B) h is the only function
    (C) f and g are the only functions
    (D) g and h are the only functions
    (E) f, g, and h are all functions

TIP

Looking for answers? All answers to exercises appear at the end of each chapter. Resist the urge to peek before trying the problems on your own.

3. What value(s) must be excluded from the domain of $f(x) = \dfrac{x+2}{x-2}$?

    (A) –2
    (B) 0
    (C) 2
    (D) 2 and –2
    (E) no value

4. If $f(x) = 3x^2 - 2x + 4$, $f(-2) =$

    (A) –12
    (B) –4
    (C) –2
    (D) 12
    (E) 20

5. If $f(x) = 4x - 5$ and $g(x) = 3^x$, then $f(g(2)) =$

    (A) 3
    (B) 9
    (C) 27
    (D) 31
    (E) none of the above

6. If $f(g(x)) = 4x^2 - 8x$ and $f(x) = x^2 - 4$, then $g(x) =$

    (A) $4 - x$
    (B) x
    (C) $2x - 2$
    (D) $4x$
    (E) x^2

7. What value(s) must be excluded from the domain of $\left(\dfrac{f}{g}\right)(x)$ if $f(x) = 3x^2 - 4x + 1$ and $g(x) = 3x^2 - 3$?

    (A) 0
    (B) 1
    (C) 3
    (D) both ±1
    (E) no values

8. If $g(x) = 3x + 2$ and $g(f(x)) = x$, then $f(2) =$

   (A) 0
   (B) 1
   (C) 2
   (D) 6
   (E) 8

9. If $p(x) = 4x - 6$ and $p(a) = 0$, then $a =$

   (A) -6

   (B) $-\dfrac{3}{2}$

   (C) $\dfrac{3}{2}$

   (D) $\dfrac{2}{3}$

   (E) 2

10. If $f(x) = e^x$ and $g(x) = \sin x$, then the value of $(f \circ g)(\sqrt{2})$ is

   (A) -0.01
   (B) -0.8
   (C) 0.34
   (D) 1.8
   (E) 2.7

11. If $f(x) = 2x - 3$, the inverse of f, f^{-1}, could be represented by

   (A) $f^{-1}(x) = 3x - 2$

   (B) $f^{-1}(x) = \dfrac{1}{2x - 3}$

   (C) $f^{-1}(x) = \dfrac{x - 2}{3}$

   (D) $f^{-1}(x) = \dfrac{x + 2}{3}$

   (E) $f^{-1}(x) = \dfrac{x + 3}{2}$

12. If $f(x) = x$, the inverse of f, f^{-1} could be represented by

   (A) $f^{-1}(x) = x$

   (B) $f^{-1}(x) = 1$

   (C) $f^{-1}(x) = \dfrac{1}{x}$

   (D) $f^{-1}(x) = y$

   (E) f^{-1} does not exist

13. The inverse of $f = \{(1,2),(2,3),(3,4),(4,1),(5,2)\}$ would be a function if the domain of f is limited to

(A) $\{1,3,5\}$
(B) $\{1,2,3,4\}$
(C) $\{1,5\}$
(D) $\{1,2,4,5\}$
(E) $\{1,2,3,4,5\}$

14. Which of the following could represent the equation of the inverse of the graph in the figure?

(A) $y = -2x + 1$

(B) $y = 2x + 1$

(C) $y = \dfrac{1}{2}x + 1$

(D) $y = \dfrac{1}{2}x - 1$

(E) $y = \dfrac{1}{2}x - \dfrac{1}{2}$

15. Which of the following relations is *even*?

  I. $y = 2$
  II. $f(x) = x$
  III. $x^2 + y^2 = 1$

(A) only I
(B) only I and II
(C) only II and III
(D) only I and III
(E) I, II, and III

16. Which of the following relations is *odd*?

  I. $y = 2$
  II. $y = x$
  III. $x^2 + y^2 = 1$

(A) only II
(B) only I and II
(C) only I and III
(D) only II and III
(E) I, II, and III

17. Which of the following relations is both *odd* and *even*?

      I. $x^2 + y^2 = 1$
     II. $x^2 - y^2 = 0$
    III. $x + y = 0$

    (A) only III
    (B) only I and II
    (C) only I and III
    (D) only II and III
    (E) I, II, and III

18. Which of the following functions is neither *odd* nor *even*?

    (A) $\{(1,2),(4,7),(-1,2),(0,4),(-4,7)\}$
    (B) $\{(1,2),(4,7),(-1,-2),(0,0),(-4,-7)\}$
    (C) $y = x^3 - 1$
    (D) $y = x^2 - 1$
    (E) $f(x) = -x$

Answers Explained

1. **(A)** Either (3,2) or (3,1), which is not an answer choice, must be removed so that 3 will be paired with only one number.

2. **(E)** For each value of x, there is only one value for y in each case. Therefore, f, g, and h are all functions.

3. **(C)** Since division by zero is forbidden, x cannot equal 2.

4. **(E)** $f(-2) = 3(-2)^2 - 2(-2) + 4 = 20$.

5. **(D)** $g(2) = 3^2 = 9$. $f(g(2)) = f(9) = 31$.

6. **(C)** To get from $f(x)$ to $f(g(x))$, x^2 must become $4x^2$. Therefore, the answer must contain $2x$ since $(2x)^2 = 4x^2$.

7. **(D)** $g(x)$ cannot equal 0. Therefore, $x \neq \pm 1$.

8. **(A)** Since $f(2)$ implies that $x = 2$, $g(f(2)) = 2$. Therefore, $g(f(2)) = 3(f(2)) + 2 = 2$. Therefore, $f(2) = 0$.

9. **(C)** $p(a) = 0$ implies $4a - 6 = 0$, so $a = \dfrac{3}{2}$.

* 10. **(E)** $(f \circ g)(\sqrt{2}) = f(g(\sqrt{2})) = f(\sin\sqrt{2}) = e^{\sin\sqrt{2}} \approx 2.7$.

11. **(E)** If $y = 2x - 3$, the inverse is $x = 2y - 3$, which is equivalent to $y = \dfrac{x+3}{2}$.

12. **(A)** Replace $f(x)$ with y ($y = x$). Interchange x and y ($x = y$), and solve for y ($y = x$), or $f^{-1}(x) = x$.

13. **(B)** The inverse is $\{(2,1),(3,2),(4,3),(1,4),(2,5)\}$, which is not a function because of (2,1) and (2,5). Choice B is the only answer choice that does not include both 1 and 5.

14. **(E)** If this line were reflected about the line $y = x$ to get its inverse, the slope would be less than 1 and the y-intercept would be less than zero. The only possibilities are Choices D and E. Choice D can be excluded because the x-intercept of $f(x)$ is greater than -1, which means the y-intercept of its inverse must be greater than -1.

15. **(D)** Use the appropriate test for determining whether a relation is even.
    I. The graph of $y = 2$ is a horizontal line, which is symmetric about the y-axis, so $y = 2$ is even.
    II. Since $f(-x) = -x \neq x = f(x)$ unless $x = 0$, this function is not even.
    III. Since $(-x)^2 + y^2 = 1$ whenever $x^2 + y^2 = 1$, this relation is even.

16. **(D)** Use the appropriate test for determining whether a relation is odd.
    I. The graph of $y = 2$ is a horizontal line, which is not symmetric about the origin, so $y = 2$ is not odd.
    II. Since $f(-x) = -x = -f(x)$, this function is odd.
    III. Since $(-x)^2 + (-y)^2 = 1$ whenever $x^2 + y^2 = 1$, this relation is odd.

17. **(B)** The analysis of relation III in the above examples indicates that I and II are both even and odd. Since $-x + y \neq 0$ when $x + y = 0$ unless $x = 0$, III is not even. Therefore, it is not both even and odd.

18. **(C)** Choice A is even, Choice B is odd, Choice D is even, and Choice E is odd. Choice C is not even because $(-x)^3 - 1 = -x^3 - 1$, which is neither $x^3 - 1$ nor $-x^3 + 1$.

Linear Functions

2

→ **DEFINITIONS**
→ **SLOPE**
→ **DISTANCE**
→ **MIDPOINT**

The equation of a **linear function** is $L(x) = mx + b$, and its graph is a line. If (x_1, y_1) and (x_2, y_2) are any two points on the line, the slope of the line is $m = \dfrac{y_2 - y_1}{x_2 - x_1}$, and the y-intercept of the graph is b, the value of y when x is zero. Vertical lines have undefined slopes and are not graphs of functions.

When given the slope m of the line and a point (x_1, y_1) on the line, a point-slope equation of the line can be written readily as $y - y_1 = m(x - x_1)$. The slope-intercept form may then be written, if required, by solving this equation for y.

The general equation of a line is $Ax + By + C = 0$. If $B \neq 0$, this equation can be solved for y: $y = -\dfrac{A}{B}x - \dfrac{C}{B}$. Thus, the slope is $m = -\dfrac{A}{B}$, and the y-intercept is $b = -\dfrac{C}{B}$. If $B = 0$ and $A \neq 0$, the general equation reduces to $x = \dfrac{C}{A}$, a vertical line.

Other important facts about nonvertical lines:

- Parallel lines have equal slopes.

- The product of the slopes of perpendicular lines is -1, or the slopes of perpendicular lines are opposite reciprocals.

- The distance between two points (x_1, y_1) and (x_2, y_2) is $\sqrt{(x_2 - x_1)^2 + (y_2 - y_1)^2}$.

- The coordinates of the midpoint of (x_1, y_1) and (x_2, y_2) is $\left(\dfrac{x_1 + x_2}{2}, \dfrac{y_1 + y_2}{2} \right)$.

- The slope of a horizontal line is 0, and its equation is $y = b$.

TIP

The standard window on most graphing calculators is defined by the intervals [-10, 10] in both directions. Because the screen is not square, perpendicular lines will look distorted. In order to achieve a visual of perpendicular lines on a graphing calculator, key ZOOM/Z-Square.

➡ Examples

(A) Describe the line $3x = 12$.

Divide both sides by 3 to get $x = 4$. This is the vertical line 4 units to the right of the y-axis.

(B) Describe the line $x - 3y - 6 = 0$.

Solve for $y = \dfrac{1}{3}x - 2$. The slope is $\dfrac{1}{3}$, and the y-intercept is -2.

(C) Write an equation of the line that contains $(6,-5)$ and has a slope of $\dfrac{3}{4}$.

In point-slope form, the equation is $y + 5 = \dfrac{3}{4}(x - 6)$.

(D) Write an equation of the line that contains (1,–3) and (–4,–2).

First find the slope $m = \dfrac{-2+3}{-4-1} = -\dfrac{1}{5}$. Then use the point $(1,-3)$ and this slope to write the

point-slope equation $y + 3 = -\dfrac{1}{5}(x - 1)$.

(E) The equation of line l_1 is $y = 2x + 3$, and the equation of line l_2 is $y = 2x - 5$.

These lines are parallel because the slope of each line is 2, and the y-intercepts are different.

(F) The equation of line l_1 is $y = \dfrac{5}{2}x - 4$, and the equation of line l_2 is $y = -\dfrac{2}{5}x + 9$.

These lines are perpendicular because the slope of l_2, $-\dfrac{2}{5}$, is the negative reciprocal of the

slope of l_1, $\dfrac{5}{2}$.

You can use these facts to write an equation of a line that is parallel or perpendicular to a given line and that contains a given point.

(G) Write an equation of the line that contains (1,7) and is parallel to the line $3x + 5y - 8 = 0$.

The slope of the given line is $-\dfrac{A}{B} = -\dfrac{3}{5}$. The point-slope equation of the line that contains

$(1,7)$ is therefore $y - 7 = -\dfrac{3}{5}(x - 1)$.

(H) Write an equation of the line that contains (–3,2) and is perpendicular to $y = 4x - 5$.

The slope of the given line is 4, so the slope of a line that is perpendicular to it is $-\dfrac{1}{4}$. The

desired equation is $y - 2 = -\dfrac{1}{4}(x + 3)$.

The distance between two points P and Q whose coordinates are (x_1,y_1) and (x_2,y_2) is given by the formula:

$$\text{Distance} = \sqrt{(x_2 - x_1)^2 + (y_2 - y_1)^2}$$

The midpoint, M, of the segment $\overline{PQ}$ has coordinates $\left(\dfrac{x_1 + x_2}{2}, \dfrac{y_1 + y_2}{2}\right)$.

(I) Given point (2,–3) and point (–5,4), find the length of $\overline{PQ}$ and the coordinates of the midpoint, M.

$$PQ = \sqrt{(-5 - 2)^2 + (4 - (-3))^2} \approx 9.9$$

$$M = \left(\dfrac{2 + (-5)}{2}, \dfrac{-3 + 4}{2}\right) = \left(\dfrac{-3}{2}, \dfrac{1}{2}\right)$$

EXERCISES

1. The slope of the line through points $A(3,-2)$ and $B(-2,-3)$ is

    (A) -5

    (B) $-\dfrac{1}{5}$

    (C) $\dfrac{1}{5}$

    (D) 1

    (E) 5

2. The slope of line $8x + 12y + 5 = 0$ is

    (A) $-\dfrac{3}{2}$

    (B) $-\dfrac{2}{3}$

    (C) $\dfrac{2}{3}$

    (D) 2

    (E) 3

3. The slope of a line that is perpendicular to line $3x - 5y + 8 = 0$ is

    (A) $-\dfrac{5}{3}$

    (B) $-\dfrac{3}{5}$

    (C) $\dfrac{3}{5}$

    (D) $\dfrac{5}{3}$

    (E) 3

4. The y-intercept of the line through the two points whose coordinates are $(1,3)$ and $(5,-2)$ is

    (A) $-\dfrac{5}{4}$

    (B) $\dfrac{5}{4}$

    (C) $\dfrac{17}{4}$

    (D) 7

    (E) 17

5. The equation of the perpendicular bisector of the segment joining the points whose coordinates are (1,4) and (−2,3) is

   (A) $3x - 2y + 5 = 0$
   (B) $x - 3y + 2 = 0$
   (C) $3x + y - 2 = 0$
   (D) $x - 3y + 11 = 0$
   (E) $x + 3y - 10 = 0$

6. The length of the segment joining the points with coordinates (−2,4) and (3,−5) is

   (A) 2.8
   (B) 3.7
   (C) 10.0
   (D) 10.3
   (E) none of the above

7. The slope of the line parallel to the line whose equation is $2x + 3y = 8$ is

   (A) −2
   (B) $-\dfrac{3}{2}$
   (C) $-\dfrac{2}{3}$
   (D) $\dfrac{2}{3}$
   (E) $\dfrac{3}{2}$

Answers Explained

1. **(C)** Slope $= \dfrac{-3-(-2)}{-2-3} = \dfrac{1}{5}$.

2. **(B)** $y = -\dfrac{2}{3}x - \dfrac{5}{12}$. The slope is $-\dfrac{2}{3}$.

3. **(A)** $y = \dfrac{3}{5}x + \dfrac{8}{5}$. The slope of the given line is $\dfrac{3}{5}$. The slope of a perpendicular line

   is $-\dfrac{5}{3}$.

4. **(C)** The slope of the line is $\dfrac{-2-3}{5-1} = -\dfrac{5}{4}$, so the point-slope equation is

   $y - 3 = -\dfrac{5}{4}(x-1)$. Solve for y to get $y = -\dfrac{5}{4}x + \dfrac{17}{4}$. The y-intercept of the line is $\dfrac{17}{4}$.

5. **(C)** The slope of the segment is $\dfrac{3-4}{-2-1} = \dfrac{1}{3}$. Therefore, the slope of a perpendicular

   line is -3. The midpoint of the segment is $\left(\dfrac{1-2}{2}, \dfrac{4+3}{2}\right) = \left(-\dfrac{1}{2}, \dfrac{7}{2}\right)$. Therefore, the

   point-slope equation is $y - \dfrac{7}{2} = -3(x + \dfrac{1}{2})$. In general form, this equation is $3x + y - 2 = 0$.

＊ 6. **(D)** Length $= \sqrt{(3+2)^2 + (-5-4)^2} \approx 10.3$.

7. **(C)** $y = -\dfrac{2}{3}x + \dfrac{8}{3}$. Therefore, the slope of the parallel line is $-\dfrac{2}{3}$.

Quadratic Functions

3

- → **DEFINITIONS**
- → **PARABOLAS**
- → **COMPLETING THE SQUARE**
- → **QUADRATIC FORMULA**
- → **DISCRIMINANT**

Quadratic functions are polynomials in which the largest exponent is $n = 2$. The standard equation of a quadratic function is $P(x) = ax^2 + bx + c$, and its graph is called a parabola.

If $a > 0$, the parabola opens up, and the lowest point on the parabola is called the vertex. If $a < 0$, the parabola opens down, and the vertex is the highest point on the parabola. The x-coordinate of the vertex is given by the formula $h = -\dfrac{b}{2a}$. Substitute h for x in the equation of the parabola to get $k = \dfrac{4ac - b^2}{4a}$, the y-coordinate of the vertex. The **axis of symmetry** of a parabola has the equation $x = h$: the graph of each side of this axis is the reflection of the other side. The equation $y = a(x - h)^2 + k$ is called the **vertex form** and is equivalent to $y = ax^2 + bx + c$.

The domain of a quadratic function is all real numbers, unless otherwise specified. If $a > 0$, the y-coordinate of the vertex is the minimum value of the quadratic function, so the range consists of all y-values greater than or equal to that coordinate. If $a < 0$, the y-coordinate of the vertex is the maximum value of the quadratic function, so the range consists of all y-values less than or equal to that coordinate.

The following examples demonstrate how to analyze the graph of a parabola.

➡ Examples

(A) Analyze the graph of $y = -x^2 + 4x + 5$.

First observe that $a = 1$, $b = 4$, $c = 5$. Since $a < 0$, the parabola opens down, and the vertex is a maximum. Since $h = -\dfrac{4}{-2} = 2$ and $k = -(2)^2 + 4(2) + 5 = 9$, the vertex is $(2,9)$, the maximum is 9, and the range is $(-\infty, 9)$. The axis of symmetry is $x = 2$, and the graph crosses the x-axis twice, at $x = -1$ and at $x = 5$. These are the two zeros of the function and these are also the solutions to the equation $-x^2 + 4x + 5 = 0$. The figure below shows the graph of $y = -x^2 + 4x + 5$.

(B) Analyze the graph of $y = 3x^2 + 2x + 5$.

First observe that $a = 3$, $b = 2$, $c = 5$. Since $a > 0$, the graph opens up, and the vertex is a minimum. Since $h = -\dfrac{2}{6} = -\dfrac{1}{3}$ and $k = 3\left(-\dfrac{1}{3}\right)^2 + 2\left(-\dfrac{1}{3}\right) + 5 = \dfrac{14}{3}$, the vertex is $\left(-\dfrac{1}{3}, \dfrac{14}{3}\right)$, the minimum is $\dfrac{14}{3}$, and the range is $\left(\dfrac{14}{3}, \infty\right)$. The axis of symmetry is $x = -\dfrac{1}{3}$. The figure below shows the graph of $y = 3x^2 + 2x + 5$. The graph does not cross the x-axis. This means that the solutions to the equation $3x^2 + 2x + 5 = 0$ are imaginary numbers.

TIP

If the zeros are complex, the parabola does not cross the *x*-axis.

As seen in Example (A), one method of solving certain quadratic equations is to transform the equation so that one side is equal to zero, factor the other side, use the Zero Product Property to equate each linear factor to zero, and then solve each linear equation.

➡ Examples

(C) Solve $x^2 - x - 12 = 0$.

Factor the left side to get $(x - 4)(x + 3) = 0$. Set each factor to zero and solve each linear equation: $x - 4 = 0$ or $x + 3 = 0$, when $x = 4$ or $x = -3$.

(D) Solve $(x - 6)^2 = 0$.

In this case, $x = 6$ twice, so $x = 6$ is said to have **multiplicity** 2.

The Quadratic Formula $x = \dfrac{-b \pm \sqrt{b^2 - 4ac}}{2a}$ can be used to obtain the exact solutions of any quadratic equation in the form $ax^2 + bx + c = 0$. This formula is the result of **completing the square** for the general quadratic.

The expression $b^2 - 4ac$ under the radical in the Quadratic Formula is called the **discriminant**. If the discriminant is positive, there are two real roots. If the discriminant is zero, there is one root of multiplicity 2. If the discriminant is negative, there is a conjugate pair of imaginary roots.

Note also that the two solutions given in the Quadratic Formula are conjugates $u + v$ and

TIP

The sum of the zeros is $-\dfrac{b}{a}$ and the product of the zeros is $\dfrac{c}{a}$.

$u - v$, where $u = -\dfrac{b}{2a}$ and $v = \dfrac{\sqrt{b^2 - 4ac}}{2a}$. This implies that the sum of the roots of a quadratic equation is $2u = 2\left(-\dfrac{b}{2a}\right) = -\dfrac{b}{a}$, while the product of the roots is:

$$u^2 - v^2 = \frac{b^2}{4a^2} - \frac{b^2 - 4ac}{4a^2} = \frac{4ac}{4a^2} = \frac{c}{a}$$

(E) Solve $2x^2 - 8x + 6 = 0$ by completing the square.

First divide both sides of the equation by 2, to make the coefficient of x^2 equal to 1: $x^2 - 4x + 3 = 0$. Next, isolate the constant term by subtracting 3 from both sides of the equation: $x^2 - 4x = -3$, and add the square of half the linear coefficient (–4) to both sides: $x^2 - 4x + 4 = 1$. The left side of the equation is now a perfect square: $(x - 2)^2 = 1$. Taking the positive and negative square roots, we get $x - 2 = 1$ or $x - 2 = -1$. Therefore, $x = 3$ or $x = 1$.

(F) Use the Quadratic Formula to solve $3x^2 - 4x + 1 = 0$.

$a = 3$, $b = -4$, and $c = 1$. Substitute these into the Quadratic Formula:

$$x = \frac{4 \pm \sqrt{16 - 12}}{6} = \frac{4 \pm 2}{6} = 1 \text{ or } \frac{1}{3}$$

(G) What is the product of the roots of $3x^2 = 6x + 4$? Transform so that one side of the equation is zero: $3x^2 - 6x - 4 = 0$.

The product of the roots is $\dfrac{c}{a} = -\dfrac{4}{3}$.

EXERCISES

1. The coordinates of the vertex of the parabola whose equation is $y = 2x^2 + 4x - 5$ are

   (A) $(2, 11)$
   (B) $(-1, -7)$
   (C) $(1, 1)$
   (D) $(-2, -5)$
   (E) $(-4, 11)$

2. The range of the function $f(x) = 5 - 4x - x^2$ is

   (A) $\{y: y \le 0\}$
   (B) $\{y: y \ge -9\}$
   (C) $\{y: y \le 9\}$
   (D) $\{y: y \ge 0\}$
   (E) $\{y: y \le 1\}$

3. The equation of the axis of symmetry of the function $y = 2x^2 + 3x - 6$ is

   (A) $x = -\dfrac{3}{2}$

   (B) $x = -\dfrac{3}{4}$

   (C) $x = -\dfrac{1}{3}$

   (D) $x = \dfrac{1}{3}$

   (E) $x = \dfrac{3}{4}$

4. Find the zeros of $y = 2x^2 + x - 6$.

    (A) 3 and 2

    (B) −3 and 2

    (C) $\dfrac{1}{2}$ and $\dfrac{3}{2}$

    (D) $-\dfrac{3}{2}$ and 1

    (E) $\dfrac{3}{2}$ and −2

5. The sum of the zeros of $y = 3x^2 - 6x - 4$ is

    (A) −2

    (B) $-\dfrac{4}{3}$

    (C) $\dfrac{4}{3}$

    (D) 2

    (E) 6

6. $x^2 + 2x + 3 = 0$ has

    (A) two real rational roots
    (B) two real irrational roots
    (C) two equal real roots
    (D) two equal rational roots
    (E) two complex conjugate roots

7. A parabola with a vertical axis has its vertex at the origin and passes through point (7,7). The parabola intersects line $y = 6$ at two points. The length of the segment joining these points is

    (A) 14
    (B) 13
    (C) 12
    (D) 8.6
    (E) 6.5

Answers Explained

1. **(B)** The x-coordinate of the vertex is $x = -\dfrac{b}{2a} = -\dfrac{4}{4} = -1$ and the y-coordinate is $y = 2(-1)^2 + 4(-1) - 5 = -7$. Hence the vertex is the point $(-1, -7)$.

2. **(C)** Find the vertex: $x = -\dfrac{b}{2a} = \dfrac{4}{-2} = -2$ and $y = 5 - 4(-2) - (-2)^2 = 9$. Since $a = -1 < 0$, the parabola opens down, so the range is $\{y : y \le 9\}$.

3. **(B)** The x-coordinate of the vertex is $x = -\dfrac{b}{2a} = -\dfrac{3}{4}$. Thus, the equation of the axis of symmetry is $x = -\dfrac{3}{4}$.

4. **(E)** $2x^2 + x - 6 = (2x - 3)(x + 2) = 0$. The zeros are $\dfrac{3}{2}$ and -2.

5. **(D)** Sum of zeros $= -\dfrac{b}{a} = -\dfrac{-6}{3} = 2$.

6. **(E)** From the discriminant $b^2 - 4ac = 4 - 4 \cdot 1 \cdot 3 = -8 < 0$.

⁎ 7. **(B)** The equation of a vertical parabola with its vertex at the origin has the form $y = ax^2$. Substitute $(7,7)$ for x and y to find $a = \dfrac{1}{7}$. When $y = 6$, $x^2 = 42$. Therefore, $x = \pm\sqrt{42}$, and the segment $= 2\sqrt{42} \approx 13$.

Higher Degree Polynomials

4

→ **ANALYSIS OF GRAPHS**
→ **SUM AND DIFFERENCE OF CUBES**
→ **QUADRATIC FORM**
→ **SYNTHETIC DIVISION**
→ **SOLVING EQUATIONS GRAPHICALLY**

Graphs of polynomials of degree greater than 2 are not as readily analyzed as are linear and quadratic polynomials. All polynomials have graphs that are continuous. Loosely speaking, this means that graphs of polynomials can be drawn without lifting the pencil from the paper. If the degree n is an even number, both ends of the graph leave the coordinate system at either the top or bottom, such as depicted in the figures below.

If n is an odd number, then the ends of the graph leave the coordinate system at opposite ends, such as those depicted in the following figures.

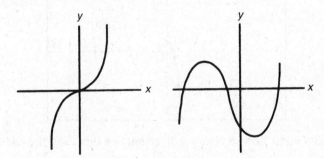

These results depict the **end behavior** of a polynomial. The continuity and end behavior polynomials make their graphs reasonably "safe" to analyze on a graphing calculator: there are no breaks or holes in the graph.

Multiplicity is the number of times a particular value appears as a root of a polynomial. Thus, for example, in the polynomial equation $f(x) = x(x - 2)^3 (x + 1)^2$, $x = 0$ has multiplicity one; $x = 2$ has multiplicity three; and $x = -1$ has multiplicity two.

The graph of *f* is shown below.

Note that when the multiplicity of a root is *odd*, the graph crosses the *x*-axis (at $x = 0, x = 2$), but when the multiplicity is *even*, the graph "kisses" the *x*-axis (at $x = -1$).

The sum of the multiplicities of a polynomial is equal to the number of roots of a polynomial. These are both equal to the degree of the polynomial. In this example, the degree of *f* is six, and the sum of the multiplicities is also six. This result is the **Fundamental Theorem of Algebra**.

➡ **Examples**_____

(A) The graph of $y = x^3 + 2x - 3$ doesn't change direction at all, and the ends go in opposite directions.

(B) The graph of $y = x^3 - 3x^2 + 1$ changes direction twice, and the ends go in opposite directions.

(C) The graph of $y = x^4 + 2x^3 - 3$ changes direction one time, and the ends go in the same direction.

(D) The graph of $y = x^4 + 3x^3 - 2x^2 - 5x + 1$ changes direction three times, and the ends go in the same direction.

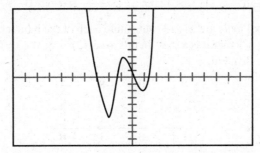

There are no general formulas for solving higher degree polynomial equations, but there are techniques for solving some of them. Sums and differences of cubes can be solved by factoring using the formulas:

$$a^3 + b^3 = (a + b)(a^2 - ab + b^2)$$

$$a^3 - b^3 = (a - b)(a^2 + ab + b^2)$$

➡ Example

(E) Solve $x^3 + 27 = 0$.

Using the formula for the sum of cubes, factor $x^3 + 27 = (x + 3)(x^2 - 3x + 9)$. By the Zero Product Property, $x + 3 = 0$ so $x = -3$, or $x^2 - 3x + 9 = 0$, which can be solved using the

Quadratic Formula: $x = \dfrac{3 \pm \sqrt{9 - 36}}{2} = \dfrac{3 \pm 3i\sqrt{3}}{2}$. The three roots are -3, $\dfrac{3 + 3i\sqrt{3}}{2}$,

and $\dfrac{3 - 3i\sqrt{3}}{2}$.

Higher degree polynomials can also take a **quadratic form**.

➡ Example

(F) Solve $x^4 - 3x^2 + 2 = 0$.

To see that this is a quadratic form, substitute $u = x^2$. The equation becomes $u^2 - 3u + 2 = 0$, which can be solved by factoring: $(u - 2)(u - 1) = 0$, so $u = 2$ or $u = 1$. Therefore, $x^2 = 2$ or $x^2 = 1$, from which it follows that $x = \pm\sqrt{2}$ or $x = \pm 1$. These are the four roots of the original equation.

Another strategy for solving higher degree polynomial equations is based on the Rational Root Theorem. This theorem states that if a rational number (the ratio of two integers $\dfrac{p}{q}$ where $q \neq 0$) is a root of a polynomial equation, then p is a divisor of the constant term and q is a divisor of the leading coefficient. Therefore, the ratios of these divisors form a list of possible rational roots of the equation. It is then a matter of substituting each of these ratios into the equation to see if any of them produces 0.

To find out whether a number r is a root, enter the polynomial into Y_1 on the graphing calculator, select "Ask" under 2nd TblSet, and enter the possible roots one at a time until $Y_1 = 0$. If all the roots are rational, they can all be found using this method. Suppose a cubic equation has exactly one rational root r. The Factor Theorem states that if r is a zero of $P(x)$, then $x - r$ is a factor. This means that the quotient $\dfrac{P(x)}{x - r}$ is a quadratic expression whose zeros are the remaining two roots of the original cubic equation. This quotient can be found by using **synthetic division**.

➥ Example _____

(G) Solve $x^3 - 8x - 3 = 0$.

The possible rational roots are $\pm 1, \pm 3$. Substitute each of the numbers in turn into the equation and determine whether it is a root. In this case, $(3)^3 - 8(3) - 3 = 0$, 3 is the only rational root. Use synthetic division:

$$
\begin{array}{r|rrrr}
3 & 1 & 0 & -8 & -3 \\
  &   & 3 & 9 & 3 \\
\hline
  & 1 & 3 & 1 & 0 \\
\end{array}
$$

Determine that $\dfrac{x^3 - 8x - 3}{x - 3} = x^2 + 3x + 1$ and then use the Quadratic Formula to conclude

that the other two roots are $\dfrac{-3 \pm \sqrt{5}}{2}$.

If none of the possible rational roots are actually roots, the equation can only be solved for real (irrational) roots by finding decimal approximations on the graphing calculator. One method is to graph the polynomial, analyze the graph as described previously to obtain a calculator window that shows all the roots (x-intercepts), and use calculator commands to obtain approximations of the zeros. The following example illustrates this technique.

➥ Example _____

(H) Solve $x^3 - 2x^2 + 2x + 4 = 0$.

Enter $x^3 - 2x^2 + 2x + 4$ into Y_1, and enter ZOOM 6. This gives you the graph of the polynomial on the standard $[-10,10]$ screen in both the x and y directions. Since the graph changes direction twice, and the end behavior is up as x gets larger and down as x gets smaller, all the important features of the graph are on the standard screen. The fact that the graph has only one x-intercept indicates that there is one real root and a conjugate pair of imaginary roots. To find the real root, enter 2nd CALC and select "zero." This command returns to the graphing screen where you scroll the cursor to the left, then the right, or the x-intercept to select lower and upper bounds. Then enter a guess (usually the right bound is fine), and the approximate zero is shown. The sequence of screens is shown below.

EXERCISES

1. $P(x) = ax^4 + x^3 - bx^2 - 4x + c$. If $P(x)$ increases without bound as x increases without bound, then, as x decreases without bound, $P(x)$

   (A) increases without bound
   (B) decreases without bound
   (C) approaches zero from above the x-axis
   (D) approaches zero from below the x-axis
   (E) cannot be determined

2. Which of the following is an odd function?

   I. $f(x) = 3x^3 + 5$
   II. $g(x) = 4x^6 + 2x^4 - 3x^2$
   III. $h(x) = 7x^5 - 8x^3 + 12x$

   (A) only I
   (B) only II
   (C) only III
   (D) only I and II
   (E) only I and III

3. How many possible rational roots are there for $2x^4 + 4x^3 - 6x^2 + 15x - 12 = 0$?

   (A) 4
   (B) 6
   (C) 8
   (D) 12
   (E) 16

4. If both $x - 1$ and $x - 2$ are factors of $x^3 - 3x^2 + 2x - 4b$, then b must be

   (A) 0
   (B) 1
   (C) 2
   (D) 3
   (E) 4

5. If $3x^3 - 9x^2 + Kx - 12$ is divisible by $x - 3$, then $K =$

   (A) −40
   (B) −3
   (C) 3
   (D) 4
   (E) 22

6. Write the equation of lowest degree with real coefficients if two of its roots are −1 and $1 + i$.

   (A) $x^3 + x^2 + 2 = 0$
   (B) $x^3 - x^2 - 2 = 0$
   (C) $x^3 - x + 2 = 0$
   (D) $x^3 - x^2 + 2 = 0$
   (E) none of the above

Answers Explained

1. **(A)** Since the degree of the polynomial is an even number, both ends of the graph go off in the same direction. Since $P(x)$ increases without bound as x increases, $P(x)$ also increases without bound as x decreases.

2. **(C)** Since the exponents are all odd, and there is no constant term, III is the only odd function.

3. **(E)** Rational roots have the form $\dfrac{p}{q}$, where p is a factor of 12 and q is a factor of 2.

   $$\frac{p}{q} \in \left\{ \pm 12, \pm 6, \pm 4, \pm 3, \pm 2, \pm 1, \pm \frac{3}{2}, \pm \frac{1}{2} \right\}.$$ The total is 16.

* 4. **(A)** Since $x - 1$ is a factor, $P(1) = 1^3 - 3 \cdot 1^2 + 2 \cdot 1 - 4b = 0$. Therefore, $b = 0$.

* 5. **(D)** Substitute 3 for x, set equal to zero, and solve for K: $3(3)^3 - 9(3)^2 + 3K - 12 = 0$, so $3K = 12$ and $K = 4$.

6. **(D)** $1 - i$ is also a root. To find the equation, multiply $(x + 1)[x - (1 + i)][x - (1 - i)]$, which are the factors that produced the three roots.

Solving Polynomial Inequalities

5

→ **SOLVE BY FACTORING**
→ **ANALYSIS OF GRAPHS**

To solve polynomial inequalities, transform the inequality so that one side is zero; then solve the corresponding equation. Because of the continuity of polynomials, these solutions are the only points where the sign of the polynomial can change. Graphically, if $P(x)$ is a polynomial, the solutions to $P(x) = 0$ are the x-intercepts of the graph. If the graph actually crosses the x-axis, then the graph either goes from above the x-axis to below it, or vice versa, in which case the sign of $y = P(x)$ changes from positive to negative or negative to positive. It is also possible for the graph to just "kiss" the x-axis from either above or below. Recall that this occurs when the multiplicity of the zero is even. These are the cases where the sign of $P(x)$ does not change at a point where $P(x) = 0$.

➡ **Examples** _____

(A) **Solve the inequality $(x - 1)(x + 4)(x + 2) < 0$.**

The polynomial is already in factored form, and the x-intercepts are $1, -4,$ and -2. By inspection of the graph below, the graph lies below the x-axis when $x < 4$ or when $-2 < x < 1$.

(B) **Solve the inequality $P(x) = x^3 - 4x^2 + x + 4 > 0$.**

Use a calculator to graph $y = P(x)$ and find its (approximate) zeros: $-0.814, 1.471,$ and 3.343.

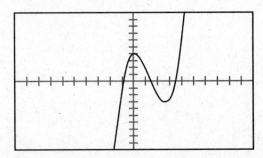

Observe that the graph of $y = P(x)$ is above the x-axis for $-0.814 < x < 1.471$ or $x > 3.343$.

(C) Solve the inequality $Q(x) = x^3 - 4x^2 + 4x > 0$.

Graph $y = Q(x)$ and factor $Q = x(x - 2)^2$ to observe that $x = 2$ with multiplicity 2, and $x = 0$.

The graph "kisses" the x-axis where a zero has even multiplicity. Therefore, $y = Q(x)$ is above the x-axis if $0 < x < 2$ or $x > 2$.

1. Which of the following is equivalent to $3x^2 - x < 2$?

   (A) $-\dfrac{3}{2} < x < 1$

   (B) $-1 < x < \dfrac{2}{3}$

   (C) $-\dfrac{2}{3} < x < 1$

   (D) $-1 < x < \dfrac{3}{2}$

   (E) $x < -\dfrac{2}{3}$ or $x > 1$

2. Solve $x^5 - 3x^3 + 2x^2 - 3 > 0$.

   (A) $(-\infty, -0.87)$
   (B) $(-1.90, -0.87)$
   (C) $(-1.90, -0.87) \cup (1.58, \infty)$
   (D) $(-0.87, 1.58)$
   (E) $(1.58, \infty)$

3. The number of integers that satisfy the inequality $x^2 + 48 < 16x$ is

   (A) 0
   (B) 4
   (C) 7
   (D) an infinite number
   (E) none of the above

4. Solve $x^2 - x - 56 \geq 0$.

   (A) $(-\infty, -7)$
   (B) $(-\infty, -7]$
   (C) $(-\infty, -7] \cup [8, \infty)$
   (D) $[8, \infty)$
   (E) $(8, \infty)$

5. Solve $\sin x < \cos x$ on $(0, 2\pi)$.

   (A) $\left(0, \dfrac{\pi}{4}\right)$

   (B) $\left(\dfrac{5\pi}{4}, 2\pi\right)$

   (C) $\left(\dfrac{\pi}{4}, \dfrac{5\pi}{4}\right)$

   (D) $\left(0, \dfrac{\pi}{4}\right) \cup \left(\dfrac{5\pi}{4}, 2\pi\right)$

   (E) $(0, 2\pi)$

Answers Explained

1. **(C)** $3x^2 - x - 2 = (3x + 2)(x - 1) = 0$ when $x = -\dfrac{2}{3}$ or 1. Numbers between these satisfy the original inequality.

✱ 2. **(C)** Graph the function, and determine that the three zeros are $-1.90, -0.87,$ and 1.58. The parts of the graph that are above the x-axis have x-coordinates between -1.90 and -0.87 and are larger than 1.58.

3. **(C)** $x^2 - 16x + 48 = (x - 4)(x - 12) = 0$, when $x = 4$ or 12. Numbers between these satisfy the original inequality.

4. **(C)** Graph the left side of the inequality, and observe where the graph is above or on the x-axis. Alternatively, factor the left side, and solve the corresponding equation $(x - 8)(x + 7) = 0$, so $x = 8$ or $x = -7$. Plot these two points on a number line to create three regions as shown below:

Since the sign of $(x - 8)(x + 7)$ is the same for all points in a region, test a point in each region: if $x = -8$, $(x - 8)(x + 7) > 0$; if $x = 0$, $(x - 8)(x + 7) < 0$; and if $x = 9$, $(x - 8)(x + 7) > 0$. Therefore, $x \le -7$ or $x \ge 8$. [This latter method may be needed later in calculus if you are not allowed to use a calculator.]

5. **(D)** Subtract $\cos x$ from both sides, graph $\sin x - \cos x$ on $(0, 2\pi)$, and observe where the graph is below the x-axis.

Triangle Trigonometry

6

→ **SOLVING TRIANGLES**
→ **LAW OF SINES**
→ **LAW OF COSINES**
→ **AMBIGUOUS CASE**
→ **FINDING AREA**
→ **SPECIAL ANGLES**

Trigonometry is often introduced in geometry courses in terms of the ratio of the sides of a right triangle. The right triangle in the figure below has right angle C and acute angles A and B. The sides opposite these angles are the hypotenuse c and legs b and a.

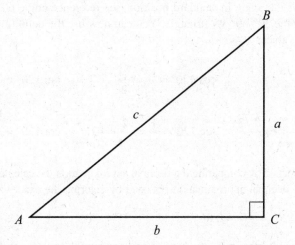

The trigonometric ratios of one of the acute angles (say A) are defined as the following:

$$\sin A = \frac{opposite\ side}{hypotenuse} = \frac{a}{c}, \cos A = \frac{adjacent\ side}{hypotenuse} = \frac{b}{c}, \text{ and } \tan A = \frac{opposite\ side}{adjacent\ side} = \frac{a}{b}$$

The mnemonic SOHCAHTOA helps students remember these ratios. It derives from **S**in is **O**pposite/**H**ypotenuse, **C**osine is **A**djacent/**H**ypotenuse, and **T**angent is **O**pposite/**A**djacent.

When you use a calculator to evaluate most trig values, you will get a decimal approximation. You can use your knowledge of the definitions of the trigonometric functions, reference angles, and the ratios of the sides of the 45°-45°-90° triangle and the 30°-60°-90° triangle ("special" triangles) to get exact trig values for "special" angles: multiples of 30°, 45°, 60°.

The ratios of the sides of the two special triangles are shown in the figures below.

45°-45°-90° Triangle

30°-60°-90° Triangle

To illustrate how this can be done, suppose you want to find the trig values of 120°. First sketch the following graph.

The graph shows the angle in standard position, the reference angle 60°, and the (signed) side length ratios for the 30°-60°-90° triangle. You can now use the definitions of the trig functions to find the trig values:

$$\sin 120° = \frac{y}{r} = \frac{\sqrt{3}}{2} \qquad \cos 120° = \frac{x}{r} = \frac{-1}{2} \qquad \tan 120° = \frac{y}{x} = \frac{\sqrt{3}}{-1} = -\sqrt{3}$$

$$\csc 120° = \frac{r}{y} = \frac{2}{\sqrt{3}} = \frac{2\sqrt{3}}{3} \qquad \sec 120° = \frac{r}{x} = \frac{2}{-1} = -2 \qquad \cot 120° = \frac{x}{y} = \frac{-1}{\sqrt{3}} = -\frac{\sqrt{3}}{3}$$

Values can be checked by comparing the decimal approximation the calculator provides for the trig function with the decimal approximation obtained by entering the exact value in a calculator.

In this example, $\sin 120° \approx \frac{\sqrt{3}}{2} \approx 0.866$.

You can also readily obtain trig values of the quadrantal angles—multiples of 90°. The terminal sides of these angles are the x- and y-axes. In these cases, you don't have a triangle at all; instead, either x or y equals 1 or –1, the other coordinate equals zero, and r equals 1. To illustrate how to use this method to evaluate the trig values of 270°, first draw the figure below.

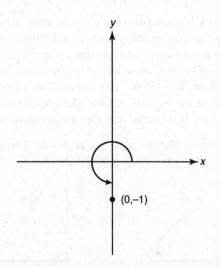

The figure indicates $x = 0$ and $y = -1$ ($r = 1$). Therefore:

$$\sin 270° = \frac{-1}{1} = -1 \qquad \cos 270° = \frac{0}{1} = 0 \qquad \tan 270° = \frac{-1}{0}, \text{ which is undefined}$$

$$\csc 270° = \frac{1}{-1} = -1 \qquad \sec 270° = \frac{1}{0}, \text{ which is undefined} \qquad \cot 270° = \frac{0}{-1} = 0$$

An arbitrary triangle has 6 parts (3 angles and 3 sides). The labeling of the angles and sides of an arbitrary triangle is the same as described for a right triangle: angles A, B, C and their opposite sides a, b, c. Typical problems in triangle trig ask you to find 3 of these values given 3 others. There are three important theorems about triangles that are based on the relationships between the trig ratios and the angles and sides of the triangle. Consider the triangle below.

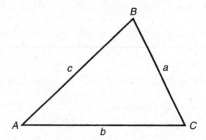

Depending on which sides and angles of the triangle are given, the following formulas can be used to find the missing parts of a triangle.

Recall from geometry that you can prove that two triangles are congruent if the three sides of one triangle are congruent to those of the other (SSS), if two sides and the included angle of one triangle are congruent to those of the other (SAS), or if two angles and one side of one triangle are congruent to those of the other (ASA or AAS). If you know three of these values for one triangle, then any other triangle that has the same three values must be congruent to the first. In other words, the other three values can be determined uniquely.

The Law of Sines and the Law of Cosines are two theorems that relate the sides and angles of a triangle.

$$\text{Law of Sines: } \frac{\sin A}{a} = \frac{\sin B}{b} = \frac{\sin C}{c}$$

The Law of Sines says that the sines of the angles of a triangle and the lengths of the opposite sides are proportional.

$$\text{Law of Cosines: } a^2 = b^2 + c^2 - 2bc \cos A$$
$$b^2 = a^2 + c^2 - 2ac \cos B$$
$$c^2 = a^2 + b^2 - 2ab \cos C$$

The Law of Cosines is a generalization of the Pythagorean theorem. Since $\cos 90° = 0$, the three forms reduce to the Pythagorean theorem with A, B, or C as the right angle. The last term of each form represents the "correction" for A, B, or C **not** being a right angle. Since the cosine of an acute angle is positive, this correction is negative if A, B, or C has a measure less than $90°$, and since the cosine of an obtuse angle is negative, the correction is positive if A, B, or C has a measure greater than $90°$.

When the measures of two angles and a side are known, the third angle can be found by subtraction, and the Law of Sines can be used to find the other two sides.

➥ Example

(A) Solve $\triangle ABC$ if $m\angle A = 45°$, $m\angle B = 57°$, $a = 4$ (AAS).

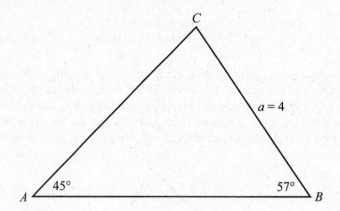

Since the measures of the three angles sum to 180° we know that $m\angle C = 78°$. By the Law of Sines: $\dfrac{\sin 45°}{4} = \dfrac{\sin 57°}{b} = \dfrac{\sin 78°}{c}$. Therefore, $b = \dfrac{4\sin 57°}{\sin 45°} \approx 4.74$ and $c = \dfrac{4\sin 78°}{\sin 45°} \approx 5.53$.

If the measures of two sides and the included angle are known, use the Law of Cosines first to find the measure of the third side. Then use the Law of Sines to find the measure of a second angle. Finally, use the fact that the sum of the angles of a triangle is 180° to find the measure of the third angle.

➥ Example

(B) Solve $\triangle ABC$ if $a = 7$, $b = 12$, $m\angle C = 62°$ (SAS).

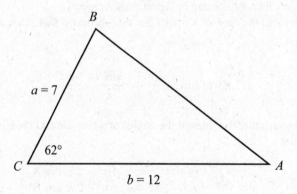

By the Law of Cosines, $c = \sqrt{7^2 + 12^2 - 2(7)(12)\cos 62°} \approx 10.68$. By the Law of Sines, $\dfrac{\sin A}{7} = \dfrac{\sin 62°}{10.68}$, so $\sin A \approx 0.5787$ and $m\angle A \approx 35.36°$. Finally:

$$m\angle C = 180° - m\angle A - m\angle B \approx 82.64$$

If the lengths of three sides of a triangle are known, use the Law of Cosines first to find the measure of one of the angles. Then use the Law of Sines to find a second angle. Finally, use the fact that the angle measures sum to 180° to find the measure of the third angle.

➠ **Example** _____

(C) Solve △ABC if $a = 3, b = 5, c = 7$ (SSS).

By the Law of Cosines: $3^2 = 5^2 + 7^2 - 2(5)(7)\cos A$, so $\cos A = \dfrac{3^2 - 5^2 - 7^2}{-2(5)(7)} \approx 0.9285$ and $m\angle A \approx 21.79°$.

By the Law of Sines, $\dfrac{\sin B}{5} = \dfrac{21.79}{3}$, so $m\angle B \approx 38.21°$.

Finally, $m\angle C \approx 180° - 21.79° - 38.21° = 120°$.

You also learned in geometry that knowing the measures of two sides and a non-included angle of one triangle are congruent to those of another triangle (SSA) does not guarantee congruence of the triangles. This SSA case is called the "ambiguous case" because there are several possible outcomes, depending on the values of the measures.

Case 1: In the figure below, $\angle A$, the given angle, is obtuse. The side with length a is the side opposite $\angle A$ and must be one of the given sides. The other given side in the figure is $\overline{AC}$, and it has length b. In the figure, $\angle A$ is obtuse. The figure shows clearly that there is no triangle unless $a > b$. In this case, the Law of Sines can be used first to find $m\angle B$. Subtract the measures of the two known angles to get $m\angle C$; then use the Law of Sines again to find c.

Solve △ABC if $m\angle A = 125°, a = 10, b = 8$.

By the Law of Sines, $\dfrac{\sin 125°}{10} = \dfrac{\sin B}{8}$. Therefore, $\sin B = \dfrac{8\sin 125°}{10} \approx 0.6553$, and

$m\angle B \approx 40.94°$. $m\angle C \approx 180° - 125° - 40.94° = 14.06°$. By the Law of Sines again,

$\dfrac{\sin 125°}{10} = \dfrac{\sin 14.06°}{c}$, so $c = \dfrac{10 \sin 14.06°}{\sin 125°} \approx 2.96$.

Case 2: If $\angle A$ is a right angle, then a is the length of the hypotenuse, and $m\angle B = \sin^{-1}\left(\dfrac{b}{a}\right)$.

Use the fact that the sum of the angles of a triangle is 180° to find the measure of the third angle, and use the Pythagorean theorem to find c, the length of the third side. This case is not likely to be seen on the Level 2 Mathematics Subject Test, but it is included here for completeness.

Solve $\triangle ABC$ if $m\angle A = 90°$, $a = 7$, $b = 5$.

$\operatorname{Sin} B = \dfrac{5}{7}$, so $m\angle B \approx 45.58$. $m\angle C \approx 90° - 45.58° = 44.42°$, and $c = \sqrt{7^2 - 5^2} \approx 4.90$.

Case 3: In the figure below, $\angle A$ is an acute angle at the bottom left, and $h = b \sin A$ is the length of the perpendicular from C to $\overline{AB}$. There are 4 subcases summarized.

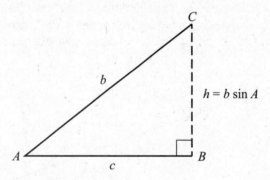

Case 3a: If $a < h$, there is no triangle.

Solve $\triangle ABC$ if $m\angle A = 50°$, $a = 3$, $b = 5$.

First calculate $h = 5 \sin 50° \approx 3.83$. Since $3 < 3.83$, there are no triangles.

Case 3b: If $a > b$, there is one triangle.

Solve $\triangle ABC$ if $m\angle A = 50°$, $a = 7$, $b = 5$.

Since $7 > 3.83$, there is one triangle, and it is obtuse. Follow the same procedure as in

Case 1 to solve this triangle: By the Law of Sines, $\dfrac{\sin 50°}{7} = \dfrac{\sin B}{5}$. Therefore,

$\sin B = \dfrac{5 \sin 50°}{7} \approx 0.5472$, and $m\angle B \approx 33.17$. Calculate the third angle as

$180° - 50° - 33.17° = 96.83°$. By the Law of Sines again, $\dfrac{\sin 50°}{7} = \dfrac{\sin 96.83}{c}$, so

$c \approx \dfrac{7 \sin 96.83°}{\sin 50°} \approx 9.07$. In this case, the triangle is obtuse.

Case 3c: If $a > h$ but $a < b$, there are two possible triangles. These are shown as $\triangle ABC$ and $\triangle AB'C$ in the figures below.

 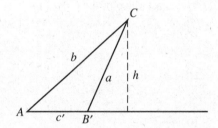

Solve $\triangle ABC$ if $m\angle A = 50°, a = 4, b = 5$.

Since $a = 4$ and $3.83 < 4 < 5$, there are two possible sides that have length 4, as shown in the figures above. This is often called "the ambiguous case" because there are **two** non-congruent triangles to solve: $\triangle ABC$ and $\triangle AB'C$. First use the Law of Sines, as in Case 3b: $\dfrac{\sin 50°}{4} = \dfrac{\sin B}{5}$, so $\sin B = \dfrac{5 \sin 50°}{4} = 0.9576$. There are **two** angles whose sine is 0.9576: $m\angle B = 73.24°$ (which is the one you get on your calculator with $\sin^{-1} 0.9576$) and $m\angle B' = 180° - 73.24° = 106.76°$. Since the sum of the angle measures in a triangle is $180°$, $m\angle ACB = 180° - 50° - 73.24° = 56.76°$ and $m\angle ACB' = 180° - 50° - 106.76° = 23.24°$.

Use the Law of Sines again to find c and c': $\dfrac{\sin 56.76°}{c} = \dfrac{\sin 50°}{4}$, so $c = 4.37$, and

$\dfrac{\sin 23.24°}{c'} = \dfrac{\sin 50°}{4}$, so $c' = 2.06$. To summarize:

$\triangle ABC$: $m\angle A = 50°, m\angle B = 73.24°, m\angle ACB = 56.76°, a = 4, b = 5, c = 4.37$

$\triangle AB'C$: $m\angle A = 50°, m\angle B' = 106.76°, m\angle ACB' = 23.24°, a = 4, b = 5, c' = 2.06$

Case 3d: It's not likely but possible that $a = h$ in which case there is one right triangle.

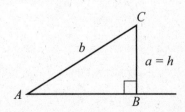

The key to remembering these cases is sketching the figures shown, especially in Case 3. Given the measures of two sides, b and c, and the included angle $\angle A$ of a triangle, there is a convenient formula for finding the **area** of $\triangle ABC$. In the figure below, the altitude to the base of measure c is $h = b \sin A$. The area of $\triangle ABC$ is, therefore, $\frac{1}{2} cb \sin A$.

Find the area of the triangle if $m\angle A = 62°$, $b = 6$, $c = 12$.

Use the formula directly to find the area of $\triangle ABC$ to be $\frac{1}{2}(12)(6)\sin 62° \approx 31.79$ units^2.

EXERCISES

1. The exact value of tan (−60°) is

   (A) $-\sqrt{3}$

   (B) -1

   (C) $-\dfrac{2}{\sqrt{3}}$

   (D) $-\dfrac{\sqrt{3}}{2}$

   (E) $-\dfrac{1}{\sqrt{3}}$

2. The exact value of $\cos \dfrac{3\pi}{4}$ is

   (A) -1

   (B) $-\dfrac{\sqrt{3}}{2}$

   (C) $-\dfrac{\sqrt{2}}{2}$

   (D) $-\dfrac{1}{2}$

   (E) 0

3. Csc 540° is

   (A) 0

   (B) $-\sqrt{3}$

   (C) $-\sqrt{2}$

   (D) -1
   (E) undefined

4. In $\triangle ABC$, $\angle A = 30°$, $b = 8$, and $a = 4\sqrt{2}$. Angle C could equal

   (A) 45°
   (B) 135°
   (C) 60°
   (D) 15°
   (E) 90°

5. In $\triangle ABC$, $\angle A = 30°$, $a = 6$, and $c = 8$. Which of the following must be true?

   (A) $0° < \angle C < 90°$
   (B) $90° < \angle C < 180°$
   (C) $45° < \angle C < 135°$
   (D) $0° < \angle C < 45°$ or $90° < \angle C < 135°$
   (E) $0° < \angle C < 45°$ or $135° < \angle C < 180°$

6. The angles of a triangle are in a ratio of 8 : 3 : 1. The ratio of the longest side of the triangle to the next longest side is

   (A) $\sqrt{6} : 2$
   (B) 8 : 3
   (C) $\sqrt{3} : 1$
   (D) 8 : 5
   (E) $2\sqrt{2} : \sqrt{3}$

7. The sides of a triangle are in a ratio of 4 : 5 : 6. The smallest angle is

   (A) 82°
   (B) 69°
   (C) 56°
   (D) 41°
   (E) 27°

8. Find the length of the longer diagonal of a parallelogram if the sides are 6 inches and 8 inches and the smaller angle is 60°.

   (A) 8
   (B) 11
   (C) 12
   (D) 7
   (E) 17

9. What are all values of side a in the figure below such that two triangles can be constructed?

(A) $a > 4\sqrt{3}$

(B) $a > 8$

(C) $a = 4\sqrt{3}$

(D) $4\sqrt{3} < a < 8$

(E) $8 < a < 8\sqrt{3}$

10. In $\triangle ABC$, $\angle B = 30°$, $\angle C = 105°$, and $b = 10$. The length of side a equals

(A) 7
(B) 9
(C) 10
(D) 14
(E) 17

11. The area of $\triangle ABC = 24\sqrt{3}$, side $a = 6$, and side $b = 16$. The $m\angle C$ is

(A) 30°
(B) 30° or 150°
(C) 60°
(D) 60° or 120°
(E) none of the above

12. The area of $\triangle ABC = 12\sqrt{3}$, side $a = 6$, and side $b = 8$. Side $c =$

(A) $2\sqrt{37}$

(B) $2\sqrt{13}$

(C) $2\sqrt{37}$ or $2\sqrt{13}$

(D) 10

(E) 10 or 12

13. Given the following data, which can form two triangles?

  I. $\angle C = 30°$, $c = 8$, $b = 12$

  II. $\angle B = 45°$, $a = 12\sqrt{2}$, $b = 15\sqrt{2}$

  III. $\angle C = 60°$, $b = 12$, $c = 5\sqrt{3}$

  (A) only I
  (B) only II
  (C) only III
  (D) only I and II
  (E) only I and III

Answers Explained

1. **(A)** Sketch a $-60°$ angle in standard position as shown in the figure below.

The tangent ratio is $\dfrac{y}{x} = \dfrac{-\sqrt{3}}{1} = -\sqrt{3}$.

2. **(C)** Sketch an angle of $\dfrac{3\pi}{4}$ radians in standard position, as shown in the figure below.

The cosine ratio is $\dfrac{x}{r} = \dfrac{-1}{\sqrt{2}} = \dfrac{-\sqrt{2}}{2}$.

3. **(E)** First, determine an angle between 0° and 360° that is coterminal with 540° by subtracting 360° from 540° repeatedly until the result is in this interval. In this case, one subtraction suffices. Since coterminal angles have the same trig values, csc 540° = csc 180°. Sketch the figure below

In a quadrantal angle $r = 1$, and the cosecant ratio is $\dfrac{r}{y}$, which is undefined.

4. **(D)** Law of Sines: $\dfrac{\sin B}{8} = \dfrac{\frac{1}{2}}{4\sqrt{2}}$. $\sin B = \dfrac{\sqrt{2}}{2}$. $B = 45°$ or $135°$. The figure below shows two possible locations for B, labeled B_1 and B_2, where $m\angle AB_1C = 45°$ and $m\angle AB_2C = 135°$. Corresponding to these, $m\angle ACB_1 = 105°$ and $m\angle ACB_2 = 15°$. Of these, only 15° is an answer choice.

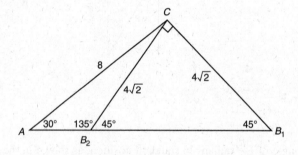

* 5. **(E)** By the Law of Sines: $\dfrac{\sin C}{8} = \dfrac{\frac{1}{2}}{6}$, so $\sin C = \dfrac{2}{3}$. The figure below shows this to be an ambiguous case (an angle, the side opposite, and another side), so $C = \sin^{-1}\left(\dfrac{2}{3}\right) \approx 41.81°$ or $C = 180° - 41.81° = 138.19°$.

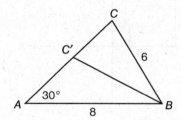

6. **(A)** The angles are $15°$, $45°$, and $120°$. Let c be the longest side and let b be the next

longest. $\dfrac{\sin 120°}{c} = \dfrac{\sin 45°}{b} \Rightarrow \dfrac{c}{b} = \dfrac{\sin 120°}{\sin 45°} = \dfrac{\dfrac{\sqrt{3}}{2}}{\dfrac{\sqrt{2}}{2}} = \dfrac{\sqrt{6}}{2}$.

✱ 7. **(D)** Use the Law of Cosines. Let the sides be 4, 5, and 6. $16 = 25 + 36 - 60 \cos A$.

Cos $A = \dfrac{45}{60} = \dfrac{3}{4}$, which implies that $A = \cos^{-1}(0.75) \approx 41°$.

✱ 8. **(C)** Law of Cosines: $d^2 = 36 + 64 - 96 \cos 120°$. $d^2 = 148$. Therefore, $d \approx 12$.

9. **(D)** Altitude to base $= 8 \sin 60° = 4\sqrt{3}$. Therefore, $4\sqrt{3} < a < 8$.

✱ 10. **(D)** $A = 45°$. Law of Sines: $\dfrac{\sin 45°}{a} = \dfrac{\sin 30°}{10}$. Therefore, $a = 10\sqrt{2} \approx 14$.

11. **(D)** Area $= \dfrac{1}{2} ab \sin C$. $24\sqrt{3} = \dfrac{1}{2} \cdot 6 \cdot 16 \sin C$. Sin $C = \dfrac{\sqrt{3}}{2}$. Therefore, $C = 60°$
or $120°$.

12. **(C)** Area $= \dfrac{1}{2} ab \sin C$. $12\sqrt{3} = \dfrac{1}{2} \cdot 6 \cdot 8 \sin C$. Sin $C = \dfrac{\sqrt{3}}{2}$. $C = 60°$ or $120°$.

Use the Law of Cosines with $60°$ and then with $120°$.

Note: At this point in the solution you know there have to be two values for C.
Therefore, the answer must be Choice C or Choice E. If $C = 10$ (from Choice E), ABC

is a right triangle with area $= \dfrac{1}{2} \cdot 6 \cdot 8 = 24$. Therefore, Choice E is not the answer,
and Choice C is the correct answer.

13. **(A)** In I, the altitude $= 12 \cdot \dfrac{1}{2} = 6$, $6 < c < 12$, and so 2 triangles. In II, $b > 12\sqrt{2}$, so only

1 triangle. In III, the altitude $= 12 \cdot \dfrac{\sqrt{3}}{2} > 5\sqrt{3}$, so no triangles.

Trigonometric Functions

7

→ **DEFINITIONS**
→ **SPECIAL ANGLES**
→ **RADIAN MEASURE**
→ **GRAPHS/AMPLITUDE/PERIOD/PHASE SHIFT**
→ **IDENTITIES**
→ **EQUATIONS**
→ **INVERSE TRIG FUNCTIONS**

The general definitions of the six trigonometric functions are obtained from an angle placed in standard position on a rectangular coordinate system. When an angle θ is placed so that its vertex is at the origin, its initial side is along the positive x-axis, and its terminal side is anywhere on the coordinate system, it is said to be in **standard position**. The angle is given a positive value if it is measured in a counterclockwise direction from the initial side to the terminal side, and a negative value if it is measured in a clockwise direction.

Let $P(x,y)$ be any point on the terminal side of the angle, and let r represent the distance between O and P. The six trigonometric functions are defined to be:

$$\sin\theta = \frac{y}{r}$$

$$\cos\theta = \frac{x}{r}$$

$$\tan\theta = \frac{y}{x}$$

$$\cot\theta = \frac{x}{y}$$

$$\sec\theta = \frac{r}{x}$$

$$\csc\theta = \frac{r}{y}$$

TIP

sin θ and cos θ are always between –1 and 1.

From these definitions it follows that:

$$\sin\theta \cdot \csc\theta = 1 \qquad \tan\theta = \frac{\sin\theta}{\cos\theta}$$

$$\cos\theta \cdot \sec\theta = 1 \qquad \cot\theta = \frac{\cos\theta}{\sin\theta}$$

$$\tan\theta \cdot \cot\theta = 1$$

If a perpendicular is constructed from P to the x-axis, a right triangle is formed by the perpendicular, the x-axis, and the terminal side. Dividing both sides of the Pythagorean theorem $x^2 + y^2 = r^2$ by r^2 yields the Pythagorean identity $\cos^2\theta + \sin^2\theta = 1$. The following identities can then be derived from this:

- $\sin^2\theta = 1 - \cos^2\theta$
- $\cos^2\theta = 1 - \sin^2\theta$
- $1 + \tan^2\theta = \sec^2\theta$
- $\cot^2\theta + 1 = \csc^2\theta$

The distance OP is always positive, and the x- and y-coordinates of P are positive or negative depending on which quadrant the terminal side of $\angle\theta$ lies in. The signs of the trigonometric functions are indicated in the following table.

Quadrant	I	II	III	IV
Function: $\sin\theta$, $\csc\theta$	+	+	–	–
$\cos\theta$, $\sec\theta$	+	–	–	+
$\tan\theta$, $\cot\theta$	+	–	+	–

TIP

All trig functions are positive in quadrant I.

S̲ine and only sine is positive in quadrant II.

T̲angent and only tangent is positive in quadrant III.

C̲osine and only cosine is positive in quadrant IV.

Just remember: A̲ll S̲tudents T̲ake C̲alculus.

Each angle θ whose terminal side lies in quadrant II, III, or IV has associated with it an angle called its **reference angle** θ_R, which is formed by the x-axis and the terminal side.

Any trig function of $\theta = \pm$ the same function of θ_R. The sign is determined by the quadrant in which the terminal side lies.

➡ Examples

(A) Express sin 320° in terms of θ_R.

$$\theta_R = 360° - 320° = 40°$$

Since the sine is negative in quadrant IV, sin 320° = –sin 40°.

(B) Express cot 200° in terms of θ_R.

$$\theta_R = 200° - 180° = 20°$$

Since the cotangent is positive in quadrant III, cot 200° = cot 20°.

(C) Express cos 130° in terms of θ_R.

$$\theta_R = 180° - 130° = 50°$$

Since the cosine is negative in quadrant II, cos 130° = –cos 50°.

Sine and cosine, tangent and cotangent, and secant and cosecant are **cofunction pairs**. **Cofunctions of complementary angles are equal.** If α and β are complementary, then trig (α) = cotrig (β) and trig (β) = cotrig (α).

➡ **Example** _____

(D) If both angles are acute and sin(3x + 20°) = cos(2x – 40°), find x.

Since these cofunctions are equal, the angles must be complementary.

$$\text{Therefore, } (3x + 20°) + (2x – 40°) = 90°$$
$$5x – 20° = 90°$$
$$x = 22°$$

Although the degree is the chief unit used to measure an angle in elementary mathematics courses, **radian measure** has several advantages in more advanced mathematics. A radian is one radius length. The circle shown in the figure below has radius r. The circumference of this circle is 360°, or 2π radians, so one radian is $\dfrac{360°}{2\pi} \approx 57.3°$.

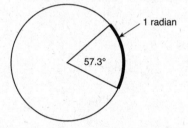

➡ **Examples** _____

In each of the following, convert the degrees to radians or the radians to degrees.
(If no unit of measurement is indicated, radians are assumed.)

(E) 30°

To change degrees to radians, multiply by $\dfrac{\pi}{180°}$, so $30° = 30°\left(\dfrac{\pi}{180°}\right) = \dfrac{\pi}{6}$.

(F) 270°

$$270°\left(\dfrac{\pi}{180°}\right) = \dfrac{3\pi}{2}$$

(G) $\dfrac{\pi}{4}$

To change radians to degrees, multiply by $\dfrac{180°}{\pi}$, so $\dfrac{\pi}{4} \cdot \dfrac{180°}{\pi} = 45°$.

(H) $\dfrac{17\pi}{3}$

$$\dfrac{17\pi}{3} \cdot \dfrac{180°}{\pi} = 1020°$$

(I) 24

$$24\left(\dfrac{180°}{\pi}\right) = \left(\dfrac{4320}{\pi}\right)° \approx 1375°$$

In a circle of radius r inches with an arc subtended by a central angle of θ measured in radians, two important formulas can be derived. The length of the arc, s, is equal to $r\theta$, and the area of the sector, AOB, is equal to $\frac{1}{2}r^2\theta$.

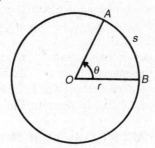

➡ Examples

(J) **Find the area of the sector and the length of the arc subtended by a central angle of $\frac{2\pi}{3}$ radians in a circle whose radius is 6 inches.**

$$s = r\theta \qquad\qquad A = \frac{1}{2}r^2\theta$$

$$s = 6 \cdot \frac{2\pi}{3} = 4\pi \text{ inches} \qquad A = \frac{1}{2} \cdot 36 \cdot \frac{2\pi}{3} = 12\pi \text{ square inches}$$

(K) **In a circle of radius 8 inches, find the area of the sector whose arc length is 6π inches.**

$$s = r\theta \qquad A = \frac{1}{2}r^2\theta$$

$$6\pi = 8\theta \qquad A = \frac{1}{2} \cdot 64 \cdot \frac{3\pi}{4} = 24\pi \text{ square inches}$$

$$\theta = \frac{3\pi}{4}$$

(L) **Find the length of the radius of a circle in which a central angle of 60° subtends an arc of length 8π inches.**

The 60° angle must be converted to radians:

$$60° = 60°\left(\frac{\pi}{180°}\right) \text{ radians} = \frac{\pi}{3} \text{ radians}$$

Therefore:

$$s = r\theta$$

$$8\pi = r \cdot \frac{\pi}{3}$$

$$r = 24 \text{ inches}$$

Analyzing the graph of a trigonometric function can be readily accomplished with the aid of a graphing calculator. Such an analysis can determine the amplitude, maximum, minimum, period, or phase shift of a trig function, or solve a trig equation or inequality.

The examples and exercises in this chapter show how a variety of trig problems can be solved without using a graphing calculator. They also explain how to solve trig equations and inequalities and how to analyze inverse trig functions.

Since the values of all the trigonometric functions repeat themselves at regular intervals, and, for some number p, $f(x) = f(x + p)$ for all numbers x, these functions are called **periodic functions**. The smallest positive value of p for which this property holds is called the **period** of the function.

The sine, cosine, secant, and cosecant have periods of 2π, and the tangent and cotangent have periods of π. The graphs of the six trigonometric functions, shown below, demonstrate that the tangent and cotangent repeat on intervals of length π and that the others repeat on intervals of length 2π.

The domain and range of each of the six trigonometric functions are summarized below.

PARENT TRIG FUNCTION

	DOMAIN	RANGE
sine	$(-\infty, \infty)$	$-1 \leq \sin x \leq 1$
cosine	$(-\infty, \infty)$	$-1 \leq \cos x \leq 1$
tangent	all real numbers except odd multiples of $\dfrac{\pi}{2}$	all real numbers
cotangent	all real numbers except all multiples of π	all real numbers
secant	all real numbers except odd multiples of $\dfrac{\pi}{2}$	$(-\infty, -1] \cup [1, \infty)$
cosecant	all real numbers except all multiples of π	$(-\infty, -1] \cup [1, \infty)$

TIP

The frequency of a trig function is the reciprocal of its period. Graphs of the parent trig functions follow.

Trig functions can be transformed just like any other function. They can be translated (slid) horizontally or vertically or dilated (stretched or shrunk) horizontally or vertically. The general form of a trigonometric function is $y = A \cdot \text{trig}(Bx + C) + D$, where trig stands for sin, cos, tan, csc, sec, or cot. The parameters A and D accomplish vertical translation and dilation, while B and C accomplish horizontal translation and dilation. When working with trig functions, the vertical dilation results in the **amplitude**, whose value is $|A|$. If B is factored out of $Bx + C$ we get $B\left(x + \dfrac{C}{B}\right)$. The horizontal translation is $-\dfrac{C}{B}$ and is called the **phase shift**, and the horizontal dilation of trig functions is measured as the **period**, which is the period of the parent trig function divided by B. Finally, D is the amount of vertical translation.

TIP

Note: sin x is an odd function.

$y = \cos x$

$y = \tan x$

$y = \cot x$

$y = \csc x$

$y = \sec x$

➡ **Examples** _____

(M) Determine the amplitude, period, and phase shift of $y = 2\sin 2x$ and sketch at least one period of the graph.

$$A = 2, B = 2, C = 0, D = 0$$

Amplitude = 2   Period = $\dfrac{2\pi}{2} = \pi$ Phase shift = 0   Vertical translation = 0

Since the phase shift is zero, the sine graph starts at its normal position, (0,0), and is drawn out to the right and to the left.

$y = 2\sin 2x$

(N) **Determine the amplitude, period, and phase shift of $y = \dfrac{1}{2}\cos\left(\dfrac{1}{2}x - \dfrac{\pi}{3}\right)$ and sketch at least one period of the graph.**

Although a graphing calculator can be used to determine the amplitude, period, and phase shift of a periodic function, it may be more efficient to derive them directly from the equation.

$$A = \frac{1}{2}, \; B = \frac{1}{2}, \; C = \frac{-\pi}{3}, \; D = 0$$

$$\text{Amplitude} = \frac{1}{2} \qquad \text{Period} = \frac{2\pi}{\frac{1}{2}} = 4\pi \qquad \text{Phase shift} = \frac{\frac{\pi}{3}}{\frac{1}{2}} = \frac{2\pi}{3} \qquad \text{Vertical translation} = 0$$

Since the phase shift is $\dfrac{2\pi}{3}$, the cosine graph starts at $x = \dfrac{2\pi}{3}$ instead of $x = 0$ and one

period ends at $x = \dfrac{2\pi}{3} + 4\pi$ or $\dfrac{14\pi}{3}$.

(O) **Determine the amplitude, period, and phase shift of $y = -2\sin(\pi x + 3\pi)$ and sketch at least one period of the graph.**

$$A = -2, \; B = \pi, \; C = 3\pi, \; D = 0$$

$$\text{Amplitude} = 2 \qquad \text{Period} = \frac{2\pi}{\pi} = 2 \qquad \text{Phase shift} = -\frac{3\pi}{\pi} = -3$$

Since the phase shift is -3, the sine graph starts at $x = -3$ instead of $x = 0$, and one period ends at $-3 + 2$ or $x = -1$. The graph can continue to the right and to the left for as many periods as desired. Since the coefficient of the sine is negative, the graph starts down as x increases from -3, instead of up as a normal sine graph does.

There are a few trigonometric identities you must know for the Level 2 Mathematics Subject Test.

- **Reciprocal Identities** recognize the definitional relationships:

$$\csc x = \frac{1}{\sin x} \qquad \sec x = \frac{1}{\cos x} \qquad \cot x = \frac{1}{\tan x}$$

- **Cofunction Identities** were discussed earlier. Using radian measure:

$$\sin x = \cos\left(\frac{\pi}{2} - x\right) \text{ and } \cos x = \sin\left(\frac{\pi}{2} - x\right)$$

$$\sec x = \csc\left(\frac{\pi}{2} - x\right) \text{ and } \csc x = \sec\left(\frac{\pi}{2} - x\right)$$

$$\tan x = \cot\left(\frac{\pi}{2} - x\right) \text{ and } \cot x = \tan\left(\frac{\pi}{2} - x\right)$$

- **Pythagorean Identities**

$$\sin^2 x + \cos^2 x = 1 \qquad \sec^2 x = 1 + \tan^2 x \qquad \csc^2 x = 1 + \cot^2 x$$

- **Double Angle Formulas**

$$\sin 2x = 2(\sin x)(\cos x) \qquad \cos 2x = \cos^2 x - \sin^2 x$$
$$= 2\cos^2 x - 1$$
$$= 1 - 2\sin^2 x$$

➡ **Examples**_____

(P) Given $\cos\theta = -\dfrac{2}{3}$ **and** $\dfrac{\pi}{2} < \theta < \pi$, **find** $\sin 2\theta$.

Since $\sin 2\theta = 2(\sin\theta)(\cos\theta)$, you need to determine the value of $\sin\theta$. From the figure below, you can see that $\sin\theta = \dfrac{\sqrt{5}}{3}$. Therefore, $\sin 2\theta = 2\,\dfrac{\sqrt{5}}{3}\left(-\dfrac{2}{3}\right) = -\dfrac{4}{9}\sqrt{5}$.

(Q) If $\cos 23° = z$, **find the value of** $\cos 46°$ **in terms of** z.

Since $46 = 2(23)$, a double angle formula can be used: $\cos 2A = 2\cos^2 A - 1$. Substituting $23°$ for A, $\cos 46° = \cos 2(23°) = 2\cos^2 23° - 1 = 2(\cos 23°)^2 - 1 = 2z^2 - 1$.

(R) If sin _x_ = _A_, find cos 2_x_ in terms of _A_.

Using the identity $\cos 2x = 1 - 2\sin^2 x$, you get $\cos 2x = 1 - 2A^2$.

You may be expected to solve trigonometric equations on the Level 2 Mathematics Subject Test by using your graphing calculator and getting answers that are decimal approximations. To solve any equation, enter each side of the equation into a function (Y_n), graph both functions, and find the point(s) of intersection on the indicated domain by choosing an appropriate window.

(S) Solve 2 sin _x_ + cos 2_x_ = 2 sin² _x_ − 1 for 0 ≤ _x_ ≤ 2π.

Enter $2\sin x + \cos 2x$ into Y_1 and $2\sin^2 x - 1$ into Y_2. Set Xmin = 0, Xmax = 2π, Ymin = −4, and Ymax = 4. Solutions (_x_-coordinates of intersection points) are 1.57, 3.67, and 5.76.

(T) Find values of _x_ on the interval [0,π] for which cos _x_ < sin 2_x_.

Enter each side of the inequality into a function, graph both, and find the values of _x_ where the graph of cos _x_ lies beneath the graph of sin 2_x_: $0.52 < x < 1.57$ or $2.62 < x < 3.14$.

If the graph of any trig function $f(x)$ is reflected about the line $y = x$, the graph of the inverse (relation) of that trig function is the result. Since all trig functions are periodic, graphs of their inverses are not graphs of functions. The domain of a trig function needs to be limited to one period so that range values are achieved exactly once. The inverse of the restricted sine function is $\sin^{-1}$; the inverse of the restricted cosine function is $\cos^{-1}$, and so forth.

Inverse Trig Function	Domain	Range
$\sin^{-1}$	$[-1, 1]$	$\left[-\dfrac{\pi}{2}, \dfrac{\pi}{2}\right]$
$\cos^{-1}$	$[-1, 1]$	$[0, \pi]$
$\tan^{-1}$	$(-\infty, \infty)$	$\left(-\dfrac{\pi}{2}, \dfrac{\pi}{2}\right)$
$\cot^{-1}$	$(-\infty, \infty)$	$(0, \pi)$
$\sec^{-1}$	$(-\infty, -1] \cup [1, \infty)$	$\left[0, \dfrac{\pi}{2}\right) \cup \left(\dfrac{\pi}{2}, \pi\right]$
$\csc^{-1}$	$(-\infty, -1] \cup [1, \infty)$	$\left[-\dfrac{\pi}{2}, 0\right) \cup \left(0, \dfrac{\pi}{2}\right]$

The inverse trig functions are used to represent angles with known trig values. If you know that the tangent of an angle is $\dfrac{8}{9}$, but you do not know the degree measure or radian measure of the angle, $\tan^{-1}\left(\dfrac{8}{9}\right)$ is an expression that represents the angle between $\dfrac{-\pi}{2}$ and $\dfrac{\pi}{2}$ whose tangent is $\dfrac{8}{9}$.

You can use your graphing calculator to find the degree or radian measure of an inverse trig value.

(U) Evaluate the radian measure of $\tan^{-1}\left(\dfrac{8}{9}\right)$.

Enter 2nd tan $\left(\dfrac{8}{9}\right)$ with your calculator in radian mode to get 0.73 radian.

(V) Evaluate the degree measure of $\sin^{-1} 0.8759$.

Enter 2nd sin (0.8759) with your calculator in degree mode to get 61.15°.

(W) Evaluate the degree measure of $\sec^{-1} 3.4735$.

First define $x = \sec^{-1} 3.4735$. If $\sec x = 3.4735$, then $\cos x = \dfrac{1}{3.4735}$. Therefore, enter 2nd

$\cos\left(\dfrac{1}{3.4735}\right)$ with your calculator in degree mode to get 73.27°.

If "trig" is any trigonometric function, $\text{trig}(\text{trig}^{-1} x) = x$. However, because of the range restriction on inverse trig functions, $\text{trig}^{-1}(\text{trig } x)$ need *not* equal x.

(X) Evaluate $\cos(\cos^{-1} 0.72)$.

$\cos(\cos^{-1} 0.72) = 0.72$.

(Y) Evaluate $\sin^{-1}(\sin 265°)$.

Enter 2nd $\sin^{-1}(\sin 265)$ with your calculator in degree mode to get –85. This is because –85° is in the required range [–90°,90°], and –85° has the same reference angle as 265°.

(Z) Evaluate $\sin\left(\cos^{-1}\dfrac{3}{5}\right)$.

Let $x = \cos^{-1}\dfrac{3}{5}$. Then $\cos x = \dfrac{3}{5}$ and x is in the first quadrant. See the figure below.

Use the Pythagorean identity $\sin^2 x + \cos^2 x = 1$ and the fact that x is in the first quadrant to

get $\sin x = \sqrt{1 - \cos^2 x} = \sqrt{\dfrac{16}{25}} = \dfrac{4}{5}$.

(AA) Evaluate $\cot^{-1}(-5.2418)$.

Let $x = \cot^{-1}(-5.2418)$. Then $\cot x = -5.2418$ and $\tan x = -\dfrac{1}{5.2418}$. Enter $\tan^{-1}\left(-\dfrac{1}{5.2418}\right)$, and it will return -0.1885. This value is in the range of $\tan^{-1}\left(-\dfrac{\pi}{2}, \dfrac{\pi}{2}\right)$, but it is not in the range of $\cot^{-1}(0, \pi)$, since $\cot x < 0$, x must be between $\dfrac{\pi}{2}$ and π and have the same reference angle as -0.1885. Therefore, $x = -0.1885 + \pi \approx 2.953$. This is not an issue with sine and cosecant because they have the same ranges, as do the cosine and secant. Thus, the calculator output for similar problems involving cosecant and secant would be correct.

EXERCISES

1. Express $\cos 320°$ as a function of an angle between $0°$ and $90°$.

   I. $\cos 40°$
   II. $\sin 50°$
   III. $\cos 50°$

   (A) I only
   (B) II only
   (C) III only
   (D) I and II
   (E) II and III

2. If point $P(-5, 12)$ lies on the terminal side of $\angle\theta$ in standard position, $\sin\theta =$

   (A) $-\dfrac{12}{13}$

   (B) $\dfrac{-5}{12}$

   (C) $\dfrac{-5}{13}$

   (D) $\dfrac{12}{13}$

   (E) $\dfrac{12}{5}$

3. If $\sec\theta = -\dfrac{5}{4}$ and $\sin\theta > 0$, then $\tan\theta =$

   (A) $\dfrac{4}{3}$

   (B) $\dfrac{3}{4}$

   (C) $-\dfrac{3}{4}$

   (D) $-\dfrac{4}{3}$

   (E) none of the above

4. If x is an angle in quadrant III and $\tan(x - 30°) = \cot x$, find x.

   (A) 240°
   (B) 225°
   (C) 210°
   (D) 60°
   (E) none of the above

5. If $90° < \alpha < 180°$ and $270° < \beta < 360°$, then which of the following *cannot* be true?

   (A) $\sin \alpha = \sin \beta$
   (B) $\tan \alpha = \sin \beta$
   (C) $\tan \alpha = \tan \beta$
   (D) $\sin \alpha = \cos \beta$
   (E) $\sec \alpha = \csc \beta$

6. Expressed as a function of an acute angle, $\cos 310° =$

   (A) $-\sin 50°$
   (B) $-\sin 40°$
   (C) $-\cos 50°$
   (D) $\cos 40°$
   (E) $\cos 50°$

7. An angle of 30 radians is equal to how many degrees?

   (A) $\dfrac{\pi}{30}$

   (B) $\dfrac{\pi}{6}$

   (C) $\dfrac{30}{\pi}$

   (D) $\dfrac{540}{\pi}$

   (E) $\dfrac{5,400}{\pi}$

8. If a sector of a circle has an arc length of 2π inches and an area of 6π square inches, what is the length of the radius of the circle?

   (A) 1
   (B) 2
   (C) 3
   (D) 6
   (E) 12

9. If a circle has a circumference of 16 inches, the area of a sector with a central angle of 4.7 radians is

   (A) 10
   (B) 12
   (C) 15
   (D) 25
   (E) 48

10. A central angle of 40° in a circle of radius 1 inch intercepts an arc whose length is *s*. Find *s*.

(A) 0.7
(B) 1.4
(C) 2.0
(D) 3.0
(E) 40

11. The pendulum on a clock swings through an angle of 25°, and the tip sweeps out an arc of 12 inches. How long is the pendulum?

(A) 1.67 inches
(B) 13.8 inches
(C) 27.5 inches
(D) 43.2 inches
(E) 86.4 inches

12. In the figure below, part of the graph of $y = \sin 2x$ is shown. What are the coordinates of point *P*?

(A) $\left(\dfrac{\pi}{2}, 1\right)$

(B) $(\pi, 1)$

(C) $\left(\dfrac{\pi}{4}, 1\right)$

(D) $\left(\dfrac{\pi}{2}, 2\right)$

(E) $(\pi, 2)$

13. The figure below could be a portion of the graph whose equation is

(A) $y - 1 = \sin x \cdot \cos x$
(B) $y \sec x = 1$
(C) $2y + 1 = \sin 2x$
(D) $2y + 1 = \cos 2x$
(E) $1 - 2y = \cos 2x$

14. As θ increases from $\dfrac{\pi}{4}$ to $\dfrac{5\pi}{4}$, the value of $4\cos\dfrac{1}{2}\theta$

   (A) increases, and then decreases
   (B) decreases, and then increases
   (C) decreases throughout
   (D) increases throughout
   (E) decreases, increases, and then decreases again

15. The function $f(x) = \sqrt{3}\cos x + \sin x$ has an amplitude of

   (A) 1.37
   (B) 1.73
   (C) 2
   (D) 2.73
   (E) 3.46

16. For what value of P is the period of the function $y = \dfrac{1}{3}\cos Px$ equal to $\dfrac{2\pi}{3}$?

   (A) $\dfrac{1}{3}$

   (B) $\dfrac{2}{3}$

   (C) 2

   (D) 3

   (E) 6

17. If $0 \le x \le \dfrac{\pi}{2}$, what is the maximum value of the function $f(x) = \sin\dfrac{1}{3}x$?

   (A) 0

   (B) $\dfrac{1}{3}$

   (C) $\dfrac{1}{2}$

   (D) $\dfrac{\sqrt{3}}{2}$

   (E) 1

18. If the graph in the figure below has an equation of the form $y = \sin(Mx + N)$, what is the value of N?

(A) $-\pi$

(B) -1

(C) $-\dfrac{1}{2}$

(D) $\dfrac{\pi}{2}$

(E) π

19. If $\sin x = \dfrac{5}{13}$ and $\cos x = -\dfrac{12}{13}$, find the value of $\sin 2x$.

(A) $-\dfrac{120}{169}$

(B) $-\dfrac{60}{169}$

(C) $\dfrac{60}{169}$

(D) $\dfrac{120}{169}$

(E) $\dfrac{10}{13}$

20. If $\tan A = \cot B$ and angles A and B are acute, then

(A) $A = B$
(B) $A = 90° + B$
(C) $B = 90° + A$
(D) $A + B = 90°$
(E) $A + B = 180°$

21. If $\cos x = \dfrac{\sqrt{3}}{2}$, find $\cos 2x$.

(A) -0.87
(B) -0.25
(C) 0
(D) 0.5
(E) 0.75

22. If $\sin 37° = z$, express $\sin 74°$ in terms of z.

    (A) $2z\sqrt{1-z^2}$

    (B) $2z^2 + 1$
    (C) $2z$
    (D) $2z^2 - 1$

    (E) $\dfrac{z}{\sqrt{1-z^2}}$

23. If $\sin x = -0.6427$, what is $\csc x$?

    (A) -1.64
    (B) -1.56
    (C) 0.64
    (D) 1.56
    (E) 1.70

24. For what value(s) of x, $0 < x < \dfrac{\pi}{2}$, is $\sin x < \cos x$?

    (A) $x < 0.79$
    (B) $x < 0.52$
    (C) $0.52 < x < 0.79$
    (D) $x > 0.52$
    (E) $x > 0.79$

25. What is the range of the function $f(x) = 5 - 6\sin(\pi x + 1)$?

    (A) $[-6,6]$
    (B) $[-5,5]$
    (C) $[-1,1]$
    (D) $[-1,11]$
    (E) $[-11,1]$

26. Find the number of degrees in $\sin^{-1}\dfrac{\sqrt{2}}{2}$.

    (A) -45
    (B) -22.5
    (C) 0
    (D) 22.5
    (E) 45

27. Find the number of radians in $\cos^{-1}(-0.5624)$.

    (A) -0.97
    (B) 0.97
    (C) 1.77
    (D) 2.17
    (E) none of the above

28. Evaluate $\tan^{-1}(\tan 128°)$.

    (A) $-128°$
    (B) $-52°$
    (C) $52°$
    (D) $128°$
    (E) none of the above

29. Which of the following is (are) true?

   I. $\sin^{-1} 1 + \sin^{-1}(-1) = 0$
   II. $\cos^{-1} 1 + \cos^{-1}(-1) = 0$
   III. $\cos^{-1} x = \cos^{-1}(-x)$ for all x in the domain of $\cos^{-1}$

   (A) only I
   (B) only II
   (C) only III
   (D) only I and II
   (E) only II and III

30. Which of the following is a solution of $\cos 3x = \dfrac{1}{2}$?

   (A) $60°$

   (B) $\dfrac{5\pi}{3}$

   (C) $\cos^{-1}\left(\dfrac{1}{6}\right)$

   (D) $\cos^{-1}\left(\dfrac{\sqrt{3}}{2}\right)$

   (E) $\dfrac{1}{3}\cos^{-1}\left(\dfrac{1}{2}\right)$

Answers Explained

1. **(D)** Since the reference angle is 40° and cosine in quadrant IV is positive, I is correct. Since $\sin\theta = \cos(90° - \theta)$, II is also correct.

2. **(D)** See the corresponding figure. Therefore, $\sin\theta = \dfrac{12}{13}$.

3. **(C)** Angle θ is in quadrant II since sec < 0 and sin > 0. Therefore, $\tan \theta = \dfrac{3}{-4} = -\dfrac{3}{4}$.

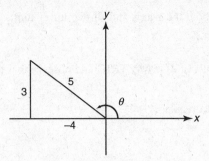

4. **(A)** Cofunctions of complementary angles are equal. $x - 30 + x = 90$ finds a reference angle of 60° for x. The angle in quadrant III that has a reference angle of 60° is 240°.

5. **(A)** Angle α is in quadrant II, and sin α is positive. Angle β is in quadrant IV, and sin β is negative.

6. **(E)** The reference angle for the fourth quarter angle of 310° is 50°. Since the cosine of a fourth quarter angle is positive, cos 310° = cos 50°.

7. **(E)** $30\left(\dfrac{180}{\pi}\right) = \dfrac{5{,}400}{\pi}$.

8. **(D)** $s = r\theta.$ $2\pi = r\theta.$ $A = \dfrac{1}{2}r^2\theta.$

$6\pi = \dfrac{1}{2}r^2\theta = \dfrac{1}{2}r(r\theta) = \dfrac{1}{2}r(2\pi).$ $r = 6.$

9. **(C)** $C = 2\pi r = 16.\, r = \dfrac{8}{\pi} \approx 2.55.$

$A = \dfrac{1}{2}r^2\theta \approx \dfrac{1}{2}(2.55)^2(4.7) \approx 15.$

✳ 10. **(A)** $40° = \dfrac{2\pi}{9} \approx 0.7.$

$s = r\theta \approx 1(0.7) \approx 0.7.$

✳ 11. **(C)** Change 25° to 0.436 radian $\left(0.436 = 25\left(\dfrac{\pi}{180}\right)\right).$

$s = r\theta$, and so $12 = r(0.436)$ and $r = 27.5$ inches.

12. **(C)** Period $= \dfrac{2\pi}{2} = \pi.$ Point P is $\dfrac{1}{4}$ of the way through the period. Amplitude is 1 because the coefficient of sin is 1. Therefore, point P is at $\left(\dfrac{\pi}{4}, 1\right).$

13. **(E)** Amplitude = $\dfrac{1}{2}$. Period = π. Graph translated $\dfrac{1}{2}$ unit up. Graph looks like a cosine graph reflected about the x-axis and shifted up $\dfrac{1}{2}$ unit.

* 14. **(C)** Graph $4\cos\left(\dfrac{1}{2}x\right)$ using ZOOM/ZTRIG and observe that the portion of the graph between $\dfrac{\pi}{4}$ and $\dfrac{5\pi}{4}$ is decreasing.

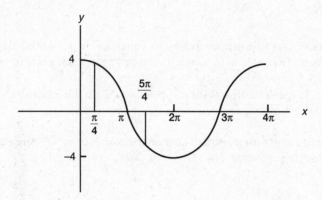

* 15. **(C)** Graph the function and determine its maximum (2) and minimum (−2). Subtract and then divide by 2.

16. **(D)** Period = $\dfrac{2\pi}{P} = \dfrac{2\pi}{3}$.

* 17. **(C)** Graph the function using 0 for Xmin and $\dfrac{\pi}{2}$ for Xmax. Observe that the maximum occurs when $x = \dfrac{\pi}{2}$. Then $f\left(\dfrac{\pi}{2}\right) = \sin\dfrac{\pi}{6} = \dfrac{1}{2}$.

18. **(D)** Period = $\dfrac{2\pi}{M} = 4\pi$ (from the figure), so $M = \dfrac{1}{2}$. Phase shift for a sine curve in the figure is $-\pi$. Therefore, $\dfrac{1}{2}x + N = 0$ when $x = -\pi$. Therefore, $N = \dfrac{\pi}{2}$.

* 19. **(A)** $\sin 2x = 2\sin x \cos x = 2\left(\dfrac{5}{13}\right)\left(-\dfrac{12}{13}\right) = -\dfrac{120}{169}$.

20. **(D)** Since tangent and cotangent are cofunctions, $\tan A = \cot(90° - A)$, so $B = 90° - A$, and $A + B = 90°$.

✱ 21. (D) $\cos 2x = 2\cos^2 x - 1 = 2\left(\dfrac{\sqrt{3}}{2}\right)^2 - 1 = \dfrac{1}{2}$ or 0.5.

22. (A) $\sin 74° = 2\sin 37° \cos 37°$. Since $\sin^2 z + \cos^2 z = 1$, $\cos z = \pm\sqrt{1 - \sin^2 z}$, and

74° is in the first quadrant, the positive square root applies, so $\cos z = \sqrt{1 - z^2}$. Therefore,

$\sin 74° = 2\sin 37°\sqrt{1 - \sin^2 37°} = 2z\sqrt{1 - z^2}$.

✱ 23. (B) $\csc x = \dfrac{1}{\sin x} = \dfrac{1}{-0.6427} = -1.56$.

✱ 24. (A) Graph $y = \sin x$ and $y = \cos x$ in radian mode using the Xmin = 0 and Xmax $= \dfrac{\pi}{2}$.

Observe that the first graph is beneath the second on [0,0.79].

25. (D) Remember that the range of the sine function is [–1,1], so the second term ranges from 6 to –6. This fact results in a range from $5 - 6 = -1$ to $5 - (-6) = 11$.

✱ 26. (E) Set your calculator to degree mode, and enter 2nd $\sin^{-1}\left(\sqrt{2}/2\right)$.

✱ 27. (D) Set your calculator to radian mode, and enter 2nd $\cos^{-1}(-0.5624)$.

✱ 28. (B) Set your calculator to degree mode, and enter 2nd $\tan^{-1}(\tan 128)$.

29. (A) Since $\sin^{-1} 1 = \dfrac{\pi}{2}$ and $\sin^{-1}(-1) = -\dfrac{\pi}{2}$, I is true. Since $\cos^{-1} 1 = 0$ and $\cos^{-1}(-1) = \pi$, II is not true. Since the range of $\cos^{-1}$ is [0,π], III is not true because $\cos^{-1}$ can never be negative.

30. (E) $3x = \cos^{-1}\left(\dfrac{1}{2}\right)$, and so $x = \dfrac{1}{3}\cos^{-1}\left(\dfrac{1}{2}\right)$.

> **Reminder:** The link to the two online practice tests can be found on the inside front cover of this book. Both online tests can be accessed on most mobile devices, including tablets and smartphones.

Exponential and Logarithmic Functions

8

→ **DEFINITIONS**
→ **PROPERTIES**
→ **ANALYSIS OF GRAPHS**

Exponential functions are of the form $y = a^x$, where $a > 0$ and $a \neq 1$. The inverse $x = a^y$ is obtained by interchanging x and y. Begin this statement with y and it reads, "y is the power of a that makes x," or $y = \log_a x$. In other words, $\log_a$ is an abbreviation for the phrase, "the power of a that makes."

The basic exponential properties:

For all positive real numbers x and y, and all real numbers a and b:

$$x^a \cdot x^b = x^{a+b} \qquad x^0 = 1$$

$$\frac{x^a}{x^b} = x^{a-b} \qquad x^{-a} = \frac{1}{x^a}$$

$$\left(x^a\right)^b = x^{ab} \qquad x^a \cdot y^a = (xy)^a$$

The basic logarithmic properties:

For all positive real numbers a, b, p, and q, and all real numbers x, where $a \neq 1$ and $b \neq 1$:

$$\log_b(p \cdot q) = \log_b p + \log_b q \qquad \log_b 1 = 0 \qquad b^{\log_b p} = p$$

$$\log_b\left(\frac{p}{q}\right) = \log_b p - \log_b q \qquad \log_b b = 1$$

$$\log_b(p^x) = x \cdot \log_b p \qquad \log_b p = \frac{\log_a p}{\log_a b} \quad \text{(Change-of-Base Formula)}$$

➥ **Examples** _____

(A) Simplify $x^{n-1} \cdot x^{2n} \cdot \left(x^{2-n}\right)^2$

This is equal to $x^{n-1} \cdot x^{2n} \cdot x^{4-2n} = x^{n-1+2n+4-2n} = x^{n+3}$.

TIP

$\log_b N$ is only defined for positive N.

(B) Simplify $\dfrac{3^{n-2} \cdot 9^{2-n}}{3^{2-n}}$.

In order to combine exponents using the properties above, the base of each factor must be the same.

$$\frac{3^{n-2} \cdot 9^{2-n}}{3^{2-n}} = \frac{3^{n-2} \cdot (3^2)^{2-n}}{3^{2-n}} = \frac{3^{n-2} \cdot 3^{4-2n}}{3^{2-n}}$$

$$= 3^{n-2+4-2n-(2-n)} = 3^0 = 1$$

(C) If log 23 = z, find the value of log 2300 in terms of z.

$$\log 2300 = \log(23 \cdot 100) = \log 23 + \log 100 = z + \log 10^2 = z + 2$$

(D) If ln 2 = x and ln 3 = y, find the value of ln 18 in terms of x and y.

$$\ln 18 = \ln(3^2 \cdot 2) = \ln 2 + 2 \ln 3 = x + 2y$$

(E) Solve for x: $\log_b(x + 5) = \log_b x + \log_b 5$.

$$\log_b x + \log_b 5 = \log_b(5x)$$

Therefore, $\log(x + 5) = \log(5x)$, which is true only when:

$$x + 5 = 5x$$
$$5 = 4x$$
$$x = \frac{5}{4}$$

(F) Evaluate $\log_{27} \sqrt{54} - \log_{27} \sqrt{6}$.

If you can enter base 27 into your calculator, enter the entire expression to get the answer $0.\bar{3} = \frac{1}{3}$. Otherwise, you can use the change-of-base formula to substitute $\frac{\log 54}{\log 27}$ for $\log_{27} \sqrt{54}$ and $\frac{\log \sqrt{6}}{\log 27}$ for $\log_{27} \sqrt{6}$. If you know your logarithms well, you can use properties to see that:

$$\log_{27} \sqrt{54} - \log_{27} \sqrt{6} = \log_{27} \sqrt{\frac{54}{6}}$$
$$= \log_{27} 3$$
$$= \frac{1}{3}$$

The graphs of all exponential functions $y = b^x$ have roughly the same shape and pass through point (0,1). If $b > 1$, the graph increases as x increases and approaches the x-axis as an asymptote as x decreases. The amount of curvature becomes greater as the value of b is made greater. If $0 < b < 1$, the graph increases as x decreases and approaches the x-axis as an asymptote as x increases. The amount of curvature becomes greater as the value of b is made closer to zero.

The graphs of all logarithmic functions $y = \log_b x$ have roughly the same shape and pass through point (1,0). If $b > 1$, the graph increases as x increases and approaches the y-axis as an asymptote as x approaches zero. The amount of curvature becomes greater as the value of b is made greater. If $0 < b < 1$, the graph decreases as x increases and approaches the y-axis as an asymptote as x approaches zero. The amount of curvature becomes greater as the value of b is made closer to zero.

Two numbers serve as special bases of exponential functions. The number 10 is convenient as a base because integer powers of ten determine place values: for instance, $10^2 = 100$ and $10^{-3} = 0.001$. The inverse of 10^x is $\log_{10} x$. By convention, the base is not written when it is 10, so the inverse of 10^x is simply written as log 10. The other special base is the number $e \approx 2.718281828....$ This is a nonterminating and nonrepeating decimal. The inverse of e^x is $\log_e x$, which is further abbreviated ln x, the **natural logarithm** of x. On the TI-84 graphing calculator, the command log will evaluate the logarithm with respect to any base by entering the value whose log is being sought followed by the base. Log base e (ln) can also be entered directly.

Exponential growth is an important application of exponential functions. Exponential growth reflects a constant rate of change from one time period to the next.

➡ **Examples**_____

(G) The population of a certain city grows at 7% per year, and the current population is 100,000. What will the population of that city be in 10 years?

Each year, the population will grow by a factor of 1.07. Therefore, after 10 years the population will be $100{,}000(1.07)^{10} \approx 196{,}715$.

(H) A bond pays interest at a rate of 3% compounded annually. How long will it take an initial investment to double?

Each year, the investment grows by a factor of 1.03. Therefore, the growth factor after n years is 1.03^n. Since the investment doubles when its growth factor is 2, we need to solve the equation $1.03^n = 2$. Take the log of both sides to get $n \log 1.03 = \log 2$. It follows that $n = \dfrac{\log 2}{\log 1.03} \approx 23.45$ years.

1. If $x^a \cdot (x^{a+1})^a \cdot (x^a)^{1-a} = x^k$, then $k =$

   (A) $2a + 1$
   (B) $a + a^2$
   (C) $3a$
   (D) $3a + 1$
   (E) $a^3 + a$

2. If $\log_8 3 = x \cdot \log_2 3$, then $x =$

   (A) $\dfrac{1}{3}$

   (B) 3
   (C) 4
   (D) $\log_4 3$
   (E) $\log_8 9$

3. If $\log_{10} m = \dfrac{1}{2}$, then $\log_{10} 10m^2 =$

   (A) 2
   (B) 2.5
   (C) 3
   (D) 10.25
   (E) 100

4. If $\log_b 5 = a$, $\log_b 2.5 = c$, and $5^x = 2.5$, then $x =$

   (A) ac

   (B) $\dfrac{c}{a}$

   (C) $a + c$
   (D) $c - a$
   (E) The value of x cannot be determined from the information given.

5. If $f(x) = \log_2 x$, then $f\left(\dfrac{2}{x}\right) + f(x) =$

   (A) $\log_2\left(\dfrac{2}{x}\right) + \log_2 x$

   (B) 1

   (C) $\log_2\left(\dfrac{2 + x^2}{x}\right)$

   (D) $\log_2\left(\dfrac{2}{x}\right) \cdot \log_2 x$

   (E) 0

6. If $\ln(xy) < 0$, which of the following must be true?

    (A) $xy < 0$
    (B) $xy < 1$
    (C) $xy > 1$
    (D) $xy > 0$
    (E) none of the above

7. If $\log_2 m = \sqrt{7}$ and $\log_7 n = \sqrt{2}$, $mn =$

    (A) 1
    (B) 2
    (C) 96
    (D) 98
    (E) 103

8. $\text{Log}_7 5 =$

    (A) 1.2
    (B) 1.1
    (C) 0.9
    (D) 0.8
    (E) -0.7

9. $\left(\sqrt[3]{2}\right)\left(\sqrt[5]{4}\right)\left(\sqrt[9]{8}\right) =$

    (A) 1.9
    (B) 2.0
    (C) 2.1
    (D) 2.3
    (E) 2.5

10. If \$300 is invested at 3%, compounded annually, how long (to the nearest year) will it take for the money to increase by 50%?

    (A) 11
    (B) 12
    (C) 13
    (D) 14
    (E) 15

Answers Explained

1. **(C)** $x^a \cdot x^{a^2+a} \cdot x^{a-a^2} = x^{a+a^2+a+a-a^2} = x^{3a}$.

2. **(A)** $x = \dfrac{\log_8 3}{\log_2 3} = \dfrac{\dfrac{\log 3}{\log 8}}{\dfrac{\log 3}{\log 2}} = \dfrac{\log 2}{3\log 2} = \dfrac{1}{3}$.

3. **(A)** $\log(10m^2) = \log 10 + 2\log m = 1 + 2 \cdot \dfrac{1}{2} = 2$.

4. **(B)** $b^a = 5$, $b^c = 2.5 = 5^x$, using the relationships between logs and exponents:

    $(b^a)^x = b^{ax} = 5^x = b^c$. Therefore, $ax = c$ and $x = \dfrac{c}{a}$.

5. **(B)** $f\left(\dfrac{2}{x}\right) + f(x) = \log_2\left(\dfrac{2}{x}\right) + \log_2 x$

    $$= \log_2 2 - \log_2 x + \log_2 x = 1.$$

6. **(B)** Since ln stands for $\log_e$, and $e > 1$, $xy < 1$.

∗ 7. **(D)** Converting the log expressions to exponential expressions gives $m = 2^{\sqrt{7}}$ and

    $n = 7^{\sqrt{2}}$. Therefore, $mn = 2^{\sqrt{7}} \cdot 7^{\sqrt{2}} \approx 6.2582 \cdot 15.673 \approx 98$.

∗ 8. **(D)** $\log_7 5 = \dfrac{\log 5}{\log 7} \approx \dfrac{0.699}{0.845} \approx 0.8$.

∗ 9. **(C)** $\sqrt[3]{2}\sqrt[5]{4}\sqrt[9]{8} = 2^{1/3}4^{1/5}8^{1/9} \approx 2.1$.

∗ 10. **(D)** Each year, the investment grows by a factor of 1.03. The growth factor after n years is 1.03^n. If the investment grows by 50%, its growth factor must be 1.5. Therefore, $1.03^t = 1.5$. Then take the log of both sides to get $\log 1.5 = t\log 1.03$ so $t = \log 1.5/\log 1.03$. Use your calculator to find that, to the nearest year, $t = 14$.

Rational Functions

9

→ **DEFINITION**
→ **VERTICAL ASYMPTOTES AND HOLES**
→ **LIMITS AT INFINITY AND HORIZONTAL ASYMPTOTES**

A function is a **rational function** if it is the ratio of two polynomials: $r(x) = \dfrac{p(x)}{q(x)}$. For the Level 2 Mathematics Subject Test, you should know about the occurrence of vertical asymptotes and holes in graphs of rational functions. A **vertical asymptote** is a vertical line $x = a$, where a is a zero of $q(x)$ but is not a zero of $p(x)$. The value a is not in the domain of r, and the y-coordinate gets infinitely larger in either the positive or negative direction as x gets closer to a. This creates a **discontinuity** (break) in the graph at $x = a$.

A **hole** is a second type of discontinuity in a graph. This occurs at a value of x that makes both the numerator $p(x)$ and the denominator $q(x)$ equal to zero. The y-coordinate of a hole is the functional value after the common factor has been divided out.

➥ **Examples**

(A) Consider $y = \dfrac{x+4}{x-3}$.

As x gets closer to 3, the denominator gets closer to zero but the numerator doesn't. The graph of this function has $x = 3$ as a vertical asymptote.

(B) Consider $y = \dfrac{x+1}{x^2-1}$.

Factor the denominator to get $y = \dfrac{x+1}{(x+1)(x-1)}$. Since y is undefined when $x = 1$ or -1, these are the values that must be checked for holes and vertical asymptotes. As x gets closer to 1, the denominator gets closer to 0 but the numerator doesn't. Therefore, $x = 1$ is a vertical asymptote. If $x \neq -1$, the factor $x + 1$ in the numerator and the denominator can be divided out, so that $y = \dfrac{1}{x-1}$. In other words, the functions $y = \dfrac{x+1}{x^2-1}$ and $y = \dfrac{1}{x-1}$ are identical except when $x = -1$. If you substitute $x = -1$ into $y = \dfrac{1}{x-1}$, you get $y = -\dfrac{1}{2}$. In other words, there is a hole at the point $\left(-1, -\dfrac{1}{2}\right)$.

In general, a graphing calculator will not show a hole because there is no pixel at that point. The hole is shown in the graph above because we know by the discussion that there is a hole when $x = -1$, and we set the window to be sure that the calculator would attempt to graph that point. Never attempt to identify a hole only by looking at the graph of a rational function.

Another aspect of rational functions that may appear on the Level 2 Mathematics Subject Test concerns the behavior of $f(x)$ as x becomes infinitely larger in either the positive or negative direction (**limits at infinity**). While there are algebraic explanations of this behavior, it is more useful to analyze it in a more intuitive fashion. This is accomplished in the following examples.

➡ Examples

(C) If $r(x) = \dfrac{3x^2 - 2x + 4}{x^3 + x - 5}$, what value does $r(x)$ approach as x gets infinitely larger?

Observe that the highest powers of x "dominate" the numerator and the denominator for very large values of x. This means that the larger x is, the more the numerator looks like $3x^2$ and the more the denominator looks like x^3. In other words, as x gets larger, $r(x) \approx \dfrac{3x^2}{x^3} = \dfrac{3}{x}$, which approaches 0.

(D) If $r(x) = \dfrac{3x^2 - 2x + 4}{x^2 + x - 5}$, what value does $r(x)$ approach as x gets infinitely larger?

In this case, $r(x) \approx \dfrac{3x^2}{x^2} = 3$ as x gets larger in the positive or negative direction.

(E) If $r(x) = \dfrac{3x^3 - 2x + 4}{x^2 + x - 5}$, what value does $r(x)$ approach as x gets infinitely larger?

Again, the highest powers of x "dominate" for large values of x, so in this case:

$$r(x) \approx \dfrac{3x^3}{x^2} = 3x$$

Since this latter quantity simply gets larger in the positive direction as x gets larger, it is said that $r(x)$ approaches ∞.

(F) If $r(x) = \dfrac{3x^3 - 2x + 4}{x^2 + x - 5}$, what value does $r(x)$ approach as x gets infinitely smaller (larger in the negative direction)?

The analysis is similar to that in Example (E), except that as x gets infinitely smaller, $3x$ also gets smaller (approaches $-\infty$).

(G) If $r(x) = \dfrac{-3x^3 - 2x + 4}{x^2 + x - 5}$, what value does $r(x)$ approach as x gets infinitely larger?

In this case, $r(x) \approx -3x$, so r gets infinitely smaller as x gets infinitely larger.

(H) If $r(x) = \dfrac{3x^4 - 2x + 4}{x^2 + x - 5}$, what value does $r(x)$ approach as x gets infinitely smaller?

This time, $r(x) \approx 3x^2$ as x gets smaller, so r gets larger since $3x^2 > 0$ for all $x \neq 0$.

These examples illustrate the behavior of a rational function as x increases in the positive or negative direction. The behavior can be summarized as follows:

If $r(x) = \dfrac{p(x)}{q(x)}$, and the leading coefficients (exponents of the largest powers) of p and q are a_m and b_n, and if x approaches $\pm\infty$, then:

- If the degree of p is **less than** the degree of q, $r(x)$ approaches 0. There is a horizontal asymptote in this case: $y = 0$ (the x-axis).

- If the degree of p is **equal to** the degree of q, $r(x)$ approaches $\dfrac{a_m}{b_n}$. There is a horizontal asymptote in this case: $y = \dfrac{a_m}{b_n}$.

- If the degree of p is **greater than** the degree of q, $r(x)$ approaches $\pm\infty$. This can be analyzed by remembering that (a) the division of numbers with the same sign results in a positive number (and the division of numbers with opposite signs results in a negative number), and (b) a negative number raised to an even power is positive (and a negative number raised to an odd power is negative). There is no horizontal asymptote in this case.

TIP

Vertical asymptotes have equations that start with $x =$ while horizontal asymptotes have equations that start with $y =$.

You can also use a graphing calculator to guess or verify limits at infinity. Returning to Example (D), we are interested in determining what value $r(x) = \dfrac{3x^2 - 2x + 4}{x^2 + x - 5}$ approaches as x gets larger. Enter the expression for $r(x)$ into Y_1, and set Indpnt in TBLSET (2nd WINDOW) to Ask. Enter TABLE (2nd GRAPH), delete existing entries, enter larger and larger values of x (e.g., 10, 100, 1,000, etc.), and observe Y_1. The TABLE screen for this activity is shown below:

X	Y1
10	2.7048
100	2.9524
1000	2.995
10000	2.9995
100000	3
100000	3

X=

It should be clear from the entries for Y_1 that the expression gets closer to 3 as x gets larger.

Although they are not rational functions, expressions that involve ratios of radicals can be analyzed in the same fashion.

(I) If $f(x) = \dfrac{\sqrt{x^2 - 3x + 10}}{x - 2}$, **what value does** $f(x)$ **approach as** x **gets infinitely larger?**

As x gets larger, the numerator looks more like $\sqrt{x^2} = |x|$, and the denominator looks more like

x, so $f(x) \approx \dfrac{|x|}{x} = \dfrac{x}{x}$ for $x > 0$. Therefore, as x gets infinitely larger, $f(x)$ approaches 1.

(J) If $f(x) = \dfrac{\sqrt{x^2 - 3x + 10}}{x - 2}$, **what value does** $f(x)$ **approach as** x **gets infinitely smaller?**

In this case, the denominator is negative, so $f(x)$ approaches -1 (approaches $-\infty$). Since $|x| = -x$

when $x < 0$, $f(x) \approx \dfrac{|x|}{x} = \dfrac{-x}{x} = -1$ as x gets infinitely smaller.

EXERCISES

1. $f(x) = \dfrac{x^3 - 8}{x^4 - 16}$. What value does $f(x)$ approach as x gets closer to 2?

   (A) 0

   (B) $\dfrac{3}{8}$

   (C) $\dfrac{1}{2}$

   (D) $\dfrac{4}{7}$

   (E) This expression is undefined.

2. What does $\dfrac{5x^2 - 2}{3x^2 + 8}$ approach as x gets infinitely larger?

   (A) $-\dfrac{1}{4}$

   (B) 0

   (C) $\dfrac{3}{11}$

   (D) $\dfrac{5}{3}$

   (E) ∞

3. Which of the following is an equation of an asymptote of the graph of $y = \dfrac{3x^2 - 2x - 1}{9x^2 - 1}$?

    I. $x = \dfrac{1}{3}$

    II. $x = -\dfrac{1}{3}$

    III. $y = \dfrac{1}{3}$

(A) I only
(B) III only
(C) I and II only
(D) I and III only
(E) I, II, and III

4. When $x = 1$, $y = \dfrac{x^4 - 1}{x^3 - 1}$ is of the form $\dfrac{0}{0}$, which is not defined. What value of y will make this graph continuous when $x = 1$?

(A) -1
(B) 0
(C) 1
(D) $\dfrac{4}{3}$
(E) 4

5. Which of the following is the equation of an asymptote of the graph of $y = \dfrac{x^2 + 7}{x + 3}$?

(A) $x = -3$
(B) $x = -\dfrac{7}{3}$
(C) $y = -\sqrt{7}$
(D) $y = 0$
(E) $y = -3$

Answers Explained

1. **(B)** If you substitute 2 for x, you would get $\frac{0}{0}$. This means that there is a hole in this function when $x = 2$. Factor the numerator and the denominator of the expression, and divide out common factors to get $\frac{(x-2)(x^2+2x+4)}{(x^2+4)(x+2)(x-2)}$. Now substitute 2 for x to get $\frac{12}{32} = \frac{3}{8}$.

2. **(D)** The degree of the numerator and the denominator are the same (2), so the expression gets closer to $\frac{5}{3}$.

3. **(D)** y approaches $\frac{1}{3}$ as x gets infinitely larger or smaller, so III is correct. Factor the numerator and the denominator of the graph in the question to get $y = \frac{(3x+1)(x-1)}{(3x+1)(3x-1)}$. Since $3x + 1$ is in both the numerator and the denominator, $x = -\frac{1}{3}$ is a hole rather than an asymptote, so II is not correct. Since $3x - 1$ is only in the denominator, $x = \frac{1}{3}$ is a vertical asymptote, so I is correct.

4. **(D)** Factor the numerator and the denominator, and divide out common factors to get $\frac{(x+1)(x-1)(x^2+1)}{(x-1)(x^2+x+1)}$. When you substitute 1 into this expression, you get $\frac{4}{3}$.

5. **(A)** When $x = -3$, the denominator is 0, but the numerator isn't 0. Therefore, $x = -3$ is an asymptote of this graph.

Parametric Equations

10

→ **DEFINITIONS**
→ **ELIMINATING THE PARAMETER**

At times, it is convenient to express a relationship between x and y in terms of a third variable, usually denoted by a **parameter** t. For example, **parametric equations** $x = x(t)$, $y = y(t)$ can be used to locate a particle on the plane at various times t.

➡ **Example** _____

(A) Graph the parametric equations $\begin{cases} x = 3t + 4 \\ y = t - 5 \end{cases}$

Select MODE on your graphing calculator, and select PAR. Enter $3t + 4$ into X_{1_T} and $t - 5$ into Y_{1_T}. The standard window uses 0 for Tmin and 6.28... (2π) for Tmax along with the usual ranges for x and y. The choice of 0 for Tmin reflects the interpretation of t as "time." With the standard window, the graph looks like the figure below.

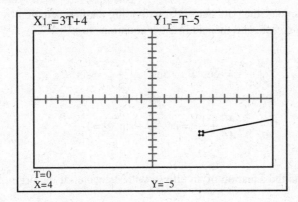

If you use TRACE, the cursor will begin at $t = 0$, where $(x,y) = (4,-5)$. As t increases from 0, the graph traces out a line that ascends as it moves right.

It may be possible to eliminate the parameter and to rewrite the equation in familiar xy-form. Just remember that the resulting equation may consist of points not on the graph of the original set of equations.

➡ **Examples** _____

(B) $\begin{cases} x = t^2 \\ y = 3t^2 + 1 \end{cases}$ **Eliminate the parameter and sketch the graph.**

Substituting x for t^2 in the second equation results in $y = 3x + 1$, which is the equation of a line with a slope of 3 and a y-intercept of 1. However, the original parametric equations indicate that $x \geq 0$ and $y \geq 1$ since t^2 cannot be negative. Thus, the proper way to indicate this set of points without the parameter is as follows: $y = 3x + 1$ and $x \geq 0$. The graph is the ray indicated in the figure below.

(C) Sketch the graph of the parametric equations $\begin{cases} x = 4\cos\theta \\ y = 3\sin\theta \end{cases}$

Replace the parameter θ with t, and enter the pair of equations. The graph has the shape of an ellipse, elongated horizontally, as shown in the figure below.

It is possible to eliminate the parameter, θ, by dividing the first equation by 4 and the second equation by 3, squaring each, and then adding the equations together.

$$\left(\frac{x}{4}\right)^2 = \cos^2\theta \quad \text{and} \quad \left(\frac{y}{3}\right)^2 = \sin^2\theta$$

$$\frac{x^2}{16} + \frac{y^2}{9} = \cos^2\theta + \sin^2\theta = 1$$

Here, $\dfrac{x^2}{16} + \dfrac{y^2}{9} = 1$ is the equation of an ellipse with its center at the origin, $a = 4$, and $b = 3$

(see Chapter 13). Since $-1 \leq \cos\theta \leq 1$ and $-1 \leq \sin\theta \leq 1$, $-4 \leq x \leq 4$ and $-3 \leq y \leq 3$ from the two parametric equations. In this case the parametric equations do not limit the graph obtained by removing the parameter.

1. In the graph of the parametric equations $\begin{cases} x = t^2 + t \\ y = t^2 - t \end{cases}$

   (A) $x \geq 0$
   (B) $x \geq -\dfrac{1}{4}$
   (C) x is any real number
   (D) $x \geq -1$
   (E) $x \leq 1$

2. The graph of $\begin{cases} x = \sin^2 t \\ y = 2\cos t \end{cases}$ is a

   (A) straight line
   (B) line segment
   (C) parabola
   (D) portion of a parabola
   (E) semicircle

3. Which of the following is (are) a pair of parametric equations that represent a circle?

   I. $\begin{cases} x = \sin\theta \\ y = \cos\theta \end{cases}$

   II. $\begin{cases} x = t \\ y = \sqrt{1 - t^2} \end{cases}$

   III. $\begin{cases} x = \sqrt{s} \\ y = \sqrt{1 - s} \end{cases}$

   (A) only I
   (B) only II
   (C) only III
   (D) only II and III
   (E) I, II, and III

4. A line has parametric equations $x = 6 + 3t$ and $y = 5 - t$, where t is the parameter. The slope of the line is

   (A) -2
   (B) -1
   (C) $-\dfrac{1}{3}$
   (D) $\dfrac{1}{5}$
   (E) 5

5. The graph of $\begin{cases} x = \sin 2x \\ y = 2\cos x \end{cases}$ is a(n)

   (A) straight line
   (B) ellipse with major axis on the y-axis
   (C) sideways figure 8
   (D) figure 8
   (E) semicircle

Answers Explained

* 1. **(B)** Graph these parametric equations for values of t between -5 and 5 and for x and y between -2.5 and 2.5.

Apparently the x-values are always greater than some value. Use the TRACE function to move the cursor as far left on the graph as it will go. This leads to a (correct) guess of $x \geq -\dfrac{1}{4}$. This can be verified by completing the square on the x equation:

$$x = \left(t^2 + t + \frac{1}{4}\right) - \frac{1}{4} = \left(t + \frac{1}{2}\right)^2 - \frac{1}{4}$$

This represents a parabola that opens up with vertex at $\left(-\dfrac{1}{2}, -\dfrac{1}{4}\right)$. Therefore, $x \geq -\dfrac{1}{4}$.

* 2. **(D)** Put your calculator in parametric mode, and enter the equations for x and y. Key ZOOM 6 to get the standard window and ZOOM to zoom in. The image is part of a sideways parabola.

* 3. **(A)** You could graph all three parametric pairs to discover that only I gives a circle. (II and III give semicircles). You can also see this by a simple analysis of the equations. Removing the parameter in I by squaring and adding gives $x^2 + y^2 = 1$, which is a circle of radius 1. Substituting x for t in the y equation of II and squaring gives $x^2 + y^2 = 1$, but $y \geq 0$, so this is only a semicircle. Squaring and substituting x^2 for s in the y equation of III gives $x^2 + y^2 = 1$, but $x \geq 0$, so this is only a semicircle.

4. **(C)** Solve the parametric equations by eliminating the parameter t. Since $t = \dfrac{x-6}{3}$:

$$y = 5 - \frac{x-6}{3} = 5 - \frac{1}{3}x + 2 = 7 - \frac{1}{3}x$$

Therefore, the slope is $-\dfrac{1}{3}$.

5. **(D)** Put your calculator in parametric mode, and enter the equations for x and y. Key ZOOM 6 to get the standard window, and observe that the graph is a figure 8.

Piecewise Functions

11

→ **DEFINITIONS**
→ **ABSOLUTE VALUE FUNCTION**
→ **POSTAGE STAMP FUNCTION**

Piecewise functions are defined by different equations on different parts of their domain. These functions are useful in modeling behavior that exhibits more than one pattern.

➡ **Example** _____

(A) Graph the function $f(x) = \begin{cases} 3 - x^2 \text{ if } x < 1 \\ x^3 - 4x \text{ if } x \geq 1 \end{cases}$

You can graph this on your graphing calculator. The 2nd TEST command displays an equal sign and five inequality signs. These return the value 1 if an X pixel satisfies the equality or inequality and 0 if it doesn't. For example, $x < 1$ will take the value 1 for pixel values of x that are less than 1 and the value 0 for pixel values of x that are greater than or equal to 1.

Enter $(3 - x^2)(x < 1) + (x^3 - 4x)(x \geq 1)$ into Y_1. For values of x less than 1, $(x - 1) = 1$ and $(x \geq 1) = 0$, so for these values only $3 - x^2$ will be graphed. The reverse is true for values of x greater than 1, so only $x^3 - 4x$ will be graphed. This graph is shown on the standard grid in the figure below.

Absolute value functions are a special type of piecewise functions. The absolute value function is defined as $|x| = \begin{cases} x \text{ if } x \geq 0 \\ -x \text{ if } x < 0 \end{cases}$

The general absolute value function has the form $f(x) = a|x - h| + k$, with the vertex at (h,k) and shaped like ∨ if $a > 0$ and like ∧ if $a < 0$. The vertex separates the two branches of the graph; h delineates the domain of all real numbers into two parts. The values are the slopes of the sides a.

The absolute value command is in the MATH/NUM menu of your graphing calculator. You can readily solve absolute value equations or inequalities by finding points of intersection.

(B) If $|x - 3| = 2$, find x.

Enter $|x - 3|$ into Y_1 and 2 into Y_2. As seen in the figure below, the points of intersection are at $x = 5$ and $x = 1$.

This is also easy to see algebraically. If $|x - 3| = 2$, then $x - 3 = 2$ or $x - 3 = -2$. Solving these equations yields the same solutions: 5 or 1. This equation also has a coordinate geometry solution: $|a - b|$ is the distance between a and b. Thus $|x - 3| = 2$ has the interpretation that x is 2 units from 3. Therefore, x must be 5 or 1.

(C) Find all values of x for which $|2x + 3| \geq 5$.

The graphical solution is shown below.

The desired values of x are on, or right and left of, the points of intersection: $x \geq 1$ or $x \leq -4$. By writing the inequality as $\left|x - \left(-\dfrac{3}{2}\right)\right| \geq \dfrac{5}{2}$, we can also interpret the solutions to the inequality as those points that are more than $2\dfrac{1}{2}$ units from $-1\dfrac{1}{2}$.

➡ **Examples**_____

(D) The graph of $f(x)$ is shown below.

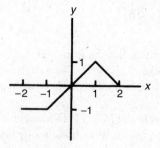

Sketch the graphs of $y|f(x)|$ and $y = f(|x|)$.

Since $|f(x)| \geq 0$, by the definition of absolute value, the graph cannot have any points below the x-axis. If $f(x) < 0$, then $|f(x)| = -f(x)$. Thus, all points below the x-axis are reflected about the x-axis, and all points above the x-axis remain unchanged.

Since the absolute value of x is taken before the function value is found, and since $|x| = -x$ when $x < 0$, any negative value of x will graph the same y-values as the corresponding positive values of x. Thus, the graph to the left of the y-axis will be a reflection of the graph to the right of the y-axis.

(E) If $f(x) = |x + 1| - 1$, what is the minimum value of $f(x)$?

Since $|x + 1| \geq 0$, its smallest value is 0. Therefore, the smallest value of $f(x)$ is $0 - 1 = -1$. The graph of $f(x)$ is indicated below.

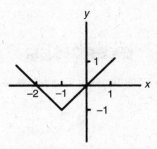

Step functions are another special type of piecewise functions. These functions are constant over different parts of their domains so that their graphs consist of horizontal segments. The greatest integer function, denoted by $[x]$, is an example of a step function. If x is an integer, then $[x] = x$. If x is not an integer, then $[x]$ is the largest integer less than x.

The greatest integer function is in the MATH/NUM menu as int on TI-83/84 calculators.

(F) Five examples of the greatest integer function integer notation are:

- [3.2] = 3
- [1.999] = 1
- [5] = 5
- [−3.12] = −4
- [−0.123] = −1

(G) Sketch the graph of $f(x) = [x]$.

TIP

On the graphing calculator, you can't tell which side of each horizontal segment has the open and closed point.

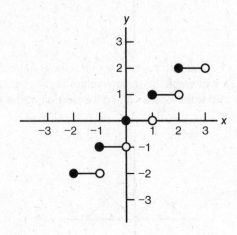

(H) What is the range of $f(x) = \left[\dfrac{[x]}{x}\right]$?

Enter int(int(x)/x) into Y₁ and choose Auto for both Indpnt and Depend in TBLSET. Set TblStart to 0 and △Tbl to 0.1. Inspection of TABLE shows only 0s and 1s as Y_1, so the range is the two-point set {0,1}.

EXERCISES

1. $|2x − 1| = 4x + 5$ has how many numbers in its solution set?

   (A) 0
   (B) 1
   (C) 2
   (D) an infinite number
   (E) none of the above

2. Which of the following is equivalent to $1 \leq |x − 2| \leq 4$?

   (A) $3 \leq x \leq 6$
   (B) $x \leq 1$ or $x \geq 3$
   (C) $1 \leq x \leq 3$
   (D) $x \leq -2$ or $x \geq 6$
   (E) $-2 \leq x \leq 1$ or $3 \leq x \leq 6$

3. The area bound by the relation $|x| + |y| = 2$ is

   (A) 8
   (B) 1
   (C) 2
   (D) 4
   (E) There is no finite area.

4. When given a function, $f(x)$, such that $f(x) = f(|x|)$, which one of the following could be the graph of $f(x)$?

(A)

(B)

(C)

(D)

(E)

5. The figure below shows the graph of which one of the following?

(A) $y = 2x - |x|$
(B) $y = |x - 1| + x$
(C) $y = |2x - 1|$
(D) $y = |x + 1| - x$
(E) $y = 2|x| - |x|$

6. The postal rate for first-class mail is 50 cents for the first ounce or portion thereof and 21 cents for each additional ounce or portion thereof up to, but not including, 3.5 ounces. The cost of a 3.5-ounce letter is $1.13. A formula for the cost in cents of first-class postage for a letter weighing N ounces ($N < 3.5$) is

(A) $50 + [N - 1] \cdot 21$
(B) $[N - 50] \cdot 21$
(C) $50 + [N] \cdot 21$
(D) $1 + [N] \cdot 21$
(E) none of the above

7. If $f(x) = n$, where n is an integer such that $n \leq x < n + 1$, the range of $f(x)$ is

(A) the set of all real numbers
(B) the set of all positive integers
(C) the set of all integers
(D) the set of all negative integers
(E) the set of all non-negative real numbers

8. If $f(x) = [4x] - 2x$ with domain $0 \leq x \leq 2$, then $f(x)$ can also be written as

(A) $2x$
(B) $-x$
(C) $-2x$
(D) $x^2 - 4x$
(E) none of the above

Answers Explained

* 1. **(B)** Enter abs($2x - 1$) into Y_1 and $4x + 5$ into Y_2. It is clear from the standard window that the two graphs intersect at only one point.

* 2. **(E)** Enter abs($x - 2$) into Y_1, 1 into Y_2, and 4 into Y_3. An inspection of the graphs shows that the values of x for which the graph of Y_1 is between the other two graphs are in two intervals. Choice E is the only answer choice that has this configuration.

* 3. **(A)** Subtract $|x|$ from both sides of the equation. Since $|y|$ cannot be negative, graph the piecewise function:

$$Y_1 = (2 - |x|)(x \geq -2 \text{ and } x \leq 2)$$
$$Y_2 = -Y_1$$

In the first command, the word "and" is in TEST/LOGIC. The result is a square that is $2\sqrt{2}$ on a side. Therefore, the area is 8.

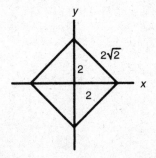

4. **(A)** Since $f(x)$ must equal $f(|x|)$, the graph must be symmetric about the y-axis. The only graph that meets this requirement is Choice A.

* 5. **(B)** Since the point where a major change takes place is at $(1,1)$, the expression in the absolute value should equal zero when $x = 1$. This occurs only in Choice B. Check your answer by graphing the function in Choice B on your graphing calculator.

6. **(E)** Choice A fails if $N = 0.5$. Choice B subtracts cents from ounces. Choice C fails if $N = 1$. Choice D adds cents to ounces.

7. **(C)** Since $f(x)$ equals an integer by definition, the answer is Choice C.

* 8. **(E)** Enter int($4x$) $- 2x$ into Y_1. The graph is shown in the figure below.

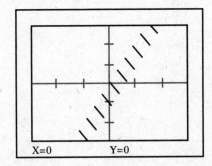

The breaks in the graph indicate that it cannot be the graph of any of the first four answer choices.

Transformations and Symmetry

12

→ **TRANSLATIONS**
→ **DILATIONS**
→ **REFLECTIONS**

The three types of transformations in the coordinate plane are translations, stretches/shrinks, and reflections. Translation preserves the shape of a graph, only moving it around the coordinate plane. Translation is accomplished by addition. Changing the scale (stretching and shrinking) can change the shape of a graph. This is accomplished by multiplication. Finally, reflection preserves the shape and size of a graph but changes its orientation. Reflection is accomplished by negation.

Suppose $y = f(x)$ defines any function.

- $y = f(x) + k$ translates $y = f(x)$ k units vertically (up if $k > 0$; down if $k < 0$)
- $y = f(x - h)$ translates $y = f(x)$ h units horizontally (right if $h > 0$; left if $h < 0$)

➡ **Examples**_____

(A) Suppose $y = f(x) = e^x$. Describe the graph of $y = e^x + 3$.

In this example, $k = 3$, so $e^x + 3 = f(x) + 3$. As shown in the figures below, each point on the graph of $y = e^x + 3$ is 3 units above the corresponding point on the graph of $y = e^x$.

(B) Suppose $y = x^2$. Describe the graph of $y = (x + 2)^2$.

In this example, $h = -2$, so $(x + 2)^2 = f(x + 2)$. As shown in the figures below, each point on the graph of $y = (x + 2)^2$ is 2 units to the left of the corresponding point on the graph of $y = x^2$.

- $y = af(x)$ stretches (shrinks) $f(x)$ vertically by a factor of $|a|$ if $|a| > 1(|a| < 1)$.

- $y = f(ax)$ shrinks (stretches) $f(x)$ horizontally by a factor of $\left|\dfrac{1}{a}\right|$ if $|a| > 1(|a| < 1)$.

(C) Suppose $y = x - 1$. Describe the graph of $y = 3(x - 1)$.

In this example, $|a| = 3$, so $3(x + 1) = 3f(x)$. As shown in the graphs below, for $x = 1.5$ the y-coordinate of each point on the graph of $y = 3(x - 1)$ is 3 times the y-coordinate of the corresponding point on the graph of $y = x - 1$.

(D) Suppose $y = x^3$. Describe the graph of $y = \left(\dfrac{1}{2}x\right)^3$.

In this example, $|a| = \dfrac{1}{2}$. As shown in the graphs below, for $y = 0.729$ the x-coordinate of each point on the graph of $y = \left(\dfrac{1}{2}x\right)^3$ is 2 times the x-coordinate of the corresponding point on the graph of $y = x^3$.

- $y = -f(x)$ reflects $y = f(x)$ about the x-axis. (The reflection is vertical.)
- $y = f(-x)$ reflects $y = f(x)$ about the y-axis. (The reflection is horizontal.)

(E) Suppose $y = \ln x$. Describe the graphs of $y = -\ln x$ and $y = \ln(-x)$.

As shown in the graphs below, the graph of $y = -\ln x$ is the reflection of the graph of $y = \ln x$ about the x-axis, and the graph of $y = \ln(-x)$ is the reflection about the y-axis.

    Translations, stretching/shrinking, and reflections can be combined to produce functions. Vertical transformations occur when adding, multiplying, or negating takes place after the function is applied (i.e., to y). The order in which multiple vertical transformations are executed does not matter. Horizontal transformations occur when adding, multiplying, or negating takes place before the function is applied (i.e., to x). Unless an order of doing transformations is specified, these transformations must be taken in the following order: reflect; change the scale; then translate. Moreover, the scale factor $|a|$ must be factored out of a translation.

(F) Suppose $y = f(x)$. Use words to describe the transformation $y = f(-ax + b)$.

Observe that all of these transformations are horizontal. First we have to write $-ax + b$ as $-a\left(x - \dfrac{b}{a}\right)$. The x-coordinate of a point (x,y) on the graph of $y = f(x)$ goes through the following sequence of transformations: reflected about the y-axis; horizontally shrunk by a factor of $|a|$ $\left(\text{or stretched by a factor of } \left|\dfrac{1}{a}\right|\right)$ and translated $\dfrac{b}{a}$ to the right.

(G) Suppose $y = \sin x$. Describe the sequence of transformations to get the graph of $y = \sin(-2x + 1)$.

Observe that all transformations are horizontal. Write the function as $y = \sin\left(-2\left(x - \dfrac{1}{2}\right)\right)$.

Consider the point $(4, -0.7568...)$ on the graph of $y = \sin x$. First reflect this point about the y-axis to $(-4, -0.7568...)$. Then shrink by a factor of 2 to $(-2, -0.7568...)$. Then translate $\dfrac{1}{2}$ units right, to $(-1.5, -0.7568...)$. The screens below show $Y_1 = \sin x$ and $Y_2 = \sin(-2x + 1)$ and a table showing the points $(4, -0.7568...)$ for Y_1 and $(-1.5, -0.7568...)$ for Y_2.

```
Plot1      Plot2       Plot3
\Y1◼sin(X)
\Y2◼sin(-2X+1)◼
\Y3=
\Y4=
\Y5=
\Y6=
\Y7=
\Y8=
\Y9=
```

X	Y₁	Y₂		
4	-.7568	-.657		
-1.5	-.9975	-.7568		

X=4

Reflections about the x- and y-axes represent two types of symmetry in a graph. Symmetry through the origin is a third type of graphical symmetry. A graph defined by an equation in x and y is symmetrical with respect to the

- x-axis if replacing x by $-x$ preserves the equation;
- y-axis if replacing y by $-y$ preserves the equation; and
- origin if replacing x and y by $-x$ and $-y$, respectively, preserves the equation.

These symmetries are defined for functions as follows:

- Symmetry about the y-axis: $f(x) = f(-x)$ for all x.
- Symmetry about the x-axis: $f(x) = -f(x)$ for all x.
- Symmetry about the origin: $f(x) = -f(-x)$ for all x.

As defined previously, functions that are symmetric about the y-axis are even functions, and those that are symmetric about the origin are odd functions.

(H) Discuss the symmetry of $f(x) = \cos x$.

Since $\cos x = \cos(-x)$, cosine is symmetric about the y-axis (an even function).

However, since $\cos x \neq -\cos x$ and $\cos(-x) \neq -\cos x$, the cosine is not symmetric with respect to the x-axis or the origin.

(I) Discuss the symmetry of $x^2 + xy + y^2 = 0$.

If you substitute $-x$ for x or $-y$ for y, but not both, the equation becomes $x^2 - xy + y^2 = 0$, which does not preserve the equation. Therefore, the graph is not symmetrical with respect to either axis. However, if you substitute both $-x$ for x and $-y$ for y, the equation is preserved, so the equation is symmetric about the origin.

EXERCISES

1. Which of the following functions transforms $y = f(x)$ by moving it 5 units to the right?

   (A) $y = f(x + 5)$
   (B) $y = f(x - 5)$
   (C) $y = f(x) + 5$
   (D) $y = f(x) - 5$
   (E) $y = 5f(x)$

2. Which of the following functions stretches $y = \cos(x)$ vertically by a factor of 3?

   (A) $y = \cos(x + 3)$

   (B) $y = \cos(3x)$

   (C) $y = \cos\left(\dfrac{1}{3}x\right)$

   (D) $y = 3 \cos x$

   (E) $y = \dfrac{1}{3}\cos x$

3. The graph of $y = f(x)$ is shown below.

   Which of the following is the graph of $y = f(-x) - 2$?

   (A)

   (B)

   (C)

   (D)

   (E)

4. Which of the following is a transformation of $y = f(x)$ that translates this function down 3, shrinks it horizontally by a factor of 2, and reflects it about the x-axis?

(A) $y = -2f(x - 3)$

(B) $y = f(-2x) - 3$

(C) $y = -f\left(\dfrac{1}{2}x\right) - 3$

(D) $y = -f(2x) - 3$

(E) $y = 2f\left(-\dfrac{1}{2}x\right) - 3$

5. Which of the following reflects the graph of $y = (x + 2)^2 - 5$ about the x-axis?

(A) $y = (x - 2)^2 + 5$

(B) $y = (2 - x)^2 - 5$

(C) $y = -(x - 2)^2 - 5$

(D) $y = (2 - x)^2 + 5$

(E) $y = -(x + 2) + 5$

6. Which of the following transforms the graph of $y = \sqrt{x}$ by translating it 3 units to the right and 2 units down and dilating it vertically by a factor of $\dfrac{1}{3}$?

(A) $y = \dfrac{1}{3}\sqrt{x + 2} - 3$

(B) $y = \dfrac{1}{3}\sqrt{x - 2} + 3$

(C) $y = \sqrt{3(x - 3)} - 2$

(D) $y = \dfrac{1}{3}\sqrt{(x - 3)} - 2$

(E) $y = \sqrt{3x - 3} - 2$

Answers Explained

1. **(B)** Horizontal translation (right) is accomplished by subtracting the amount of the translation (5) from x before the function is applied.

2. **(D)** Vertical stretching is accomplished by multiplying the function by the stretching factor after the function is applied.

3. **(D)** The graph of $y = f(-x) - 2$ reflects $y = f(x)$ about the y-axis and translates it down 2.

4. **(D)** The horizontal shrinking by a factor of 2 is the multiplication of x by 2 before the function is applied. The reflection about the x-axis is the negation of the function after it is applied. The translation down 3 is the addition of -3 after the function is applied.

5. **(E)** In order to reflect a graph about the x-axis, replace y with $-y$.

6. **(D)** A translation 3 units to the right is captured in $(x - 3)$, and a translation 2 units down is the term -2. The $\dfrac{1}{3}$ outside the $\sqrt{}$ represents the vertical dilation by a factor of $\dfrac{1}{3}$.

Conic Sections

13

→ **CIRCLES**
→ **PARABOLAS**
→ **ELLIPSES**
→ **HYPERBOLAS**

Geometrically, the conic sections are formed by the intersection of a plane and the two nappes of a right cone. The plane does not intersect the cone's vertex (see below). In a **parabola**, the plane is parallel to a lateral edge of the cone. In a **hyperbola**, the plane is parallel to the axis of the cone. In an **ellipse**, the plane is not parallel to either a lateral edge or the axis of the cone. Intersections that contain the cone's vertex are called **degenerate conics** and are not covered on the Level 2 Mathematics Subject Test.

parabola

hyperbola

ellipse

Conics are also defined in terms of points in the coordinate plane. Formulas for the distance between two points and the distance between a point and a line are then used to define standard equations of the conics. Each conic has a pair of standard equations, depending on whether the conic has an orientation to the *x*- or *y*-axis.

Each conic also has a "center." This section develops the equations of conic sections that have centers at the origin. Because of the limited orientation of conics for the Level 2 Mathematics Subject Test, equations for conics with centers at other points (h,k) are readily obtained by horizontal and vertical translation.

CIRCLES

A **circle** is a set of points that are equidistant (**radius**) from a given point (**center**). Since the center is at the origin, the standard equation of a circle is $x^2 + y^2 = r^2$, where r is the radius.

Circles have neither *x*- nor *y*-orientation. Another form of the equation of a circle is $\dfrac{x^2}{r^2} + \dfrac{y^2}{r^2} = 1$.

As you will see after reviewing the Ellipses section that follows, this form of the equation of a circle is similar to the equation of an ellipse. A circle with radius 5 is shown in the figure that follows.

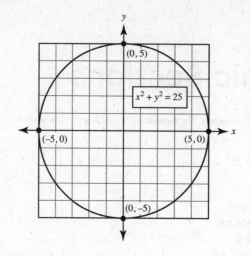

PARABOLAS

A **parabola** is the set of points that are equidistant from a given point (its **focus**) and a given line (its **directrix**). The **vertex** of a parabola is its center (at the origin). A parabola has an x-orientation if its standard equation is $y^2 = 4px$. In this case:

- The focus is at the point $(p,0)$, and the directrix has the equation $x = -p$.
- The parabola opens right if $p > 0$ and opens left if $p < 0$.

A parabola has a y-orientation if its standard equation is $x^2 = 4py$. In this case:

- The focus is at the point $(0,p)$, and the directrix has the equation $y = -p$.
- The parabola opens up if $p > 0$ and opens down if $p < 0$.

Parabolas with centers at the origin and $p = 2$ are shown in the figures below.

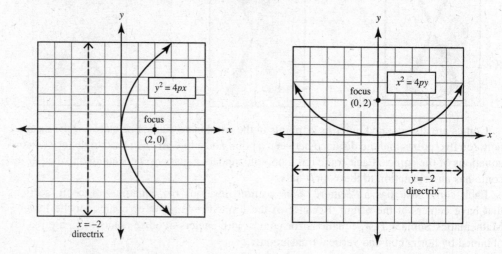

The equations of parabolas that you studied in algebra never had y^2 terms because these would not be graphs of functions. This should help you remember which parabolas have vertical orientation (those with y and not y^2 in them).

ELLIPSES

An **ellipse** is the set of points whose distances from two given points (**foci**, the plural of **focus**) *sum* to a constant. The center of an ellipse is the origin. An ellipse has an x-orientation if it is elongated horizontally. Its standard equation is $\dfrac{x^2}{a^2} + \dfrac{y^2}{b^2} = 1$ and $a^2 > b^2$. In this case:

- The **major axis** is the x-axis. It has endpoints $(-a,0)$ and $(a,0)$, and its length is $2a$.
- The **minor axis** is the y-axis. It has endpoints $(0,-b)$ and $(0,b)$, and its length is $2b$.
- The foci are at $(c,0)$ and $(-c,0)$, where $c^2 = a^2 - b^2$.

An ellipse has a y-orientation if it is elongated vertically. Its standard equation is $\dfrac{x^2}{b^2} + \dfrac{y^2}{a^2} = 1$ and $a^2 > b^2$. In this case:

- The **major axis** is the y-axis. It has endpoints $(0,-a)$ and $(0,a)$, and its length is $2a$.
- The **minor axis** is the x-axis. It has endpoints $(-b,0)$ and $(b,0)$, and its length is $2b$.
- The foci are at $(0,-c)$ and $(0,c)$, where $c^2 = a^2 - b^2$.

An ellipse has an x-orientation if a^2 (the larger of a^2 and b^2) is the denominator of x^2 and has a y-orientation if a^2 is the denominator of y^2. This provides an easy way of remembering the orientation of an ellipse. Note that an ellipse with $a^2 = b^2$ is a circle. Ellipses with the two orientations and $a = 5$, $b = 3$, and $c = 4$ are shown in the figures below.

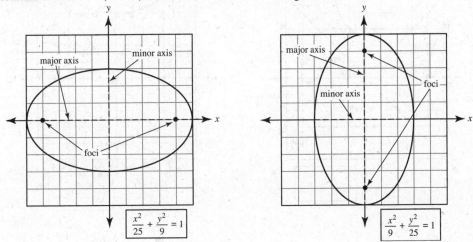

$$\frac{x^2}{25} + \frac{y^2}{9} = 1$$

$$\frac{x^2}{9} + \frac{y^2}{25} = 1$$

HYPERBOLAS

A **hyperbola** is a set of points whose distance from two given points (foci) *differ* by a constant. A hyperbola consists of two parts that correspond to the nappes of a cone. Each part has a vertex and sides that are asymptotic to a pair of intersecting lines. (These are the lines that would be formed by a plane intersecting the vertex of the cone and containing its axis.) The segment joining the two vertices is called the **transverse axis**. The conjugate axis is perpendicular to the transverse axis, and it is parallel to the segment between the two asymptotes that passes through a vertex. The center of a hyperbola is where the two axes intersect. (See the figures on page 136.)

A hyperbola has an x-orientation if its standard equation is $\dfrac{x^2}{a^2} - \dfrac{y^2}{b^2} = 1$. In this case:

- The hyperbola opens left and right.
- The vertices are $(-a,0)$ and $(a,0)$, and the transverse axis has length $2a$.
- The endpoints of the conjugate axis are $(0,-b)$ and $(0,b)$, and its length is $2b$.
- The foci are at $(-c,0)$ and $(c,0)$, where $c^2 = a^2 + b^2$.
- The equations of the asymptotes are $y = \pm\dfrac{b}{a}x$.

A hyperbola has a y-orientation if its standard equation is $\dfrac{y^2}{a^2} - \dfrac{x^2}{b^2} = 1$. In this case:

- The hyperbola opens up and down.
- The vertices are $(0,-a)$ and $(0,a)$, and the transverse axis has length $2a$.
- The endpoints of the conjugate axis are $(-b,0)$ and $(b,0)$, and its length is $2b$.
- The foci are at $(0,-c)$ and $(0,c)$, where $c^2 = a^2 + b^2$.
- The equations of the asymptotes are $y = \pm\dfrac{a}{b}x$.

Rather than memorize the formulas for the slopes of the asymptotes of a hyperbola, you can solve its standard equation after replacing 1 with 0. If the x^2 term is positive, $\dfrac{x^2}{a^2} - \dfrac{y^2}{b^2} = 0$, so $y^2 = \dfrac{b^2 x^2}{a^2}$ and $y = \pm \dfrac{b}{a} x$. If the y^2 term is positive, $\dfrac{y^2}{a^2} - \dfrac{x^2}{b^2} = 0$, so $y^2 = \dfrac{a^2 x^2}{b^2}$ and $y = \pm \dfrac{a}{b} x$. Two hyperbolas centered at the origin with $a = 4$, $b = 3$, and $c = 5$ are shown in the figures below.

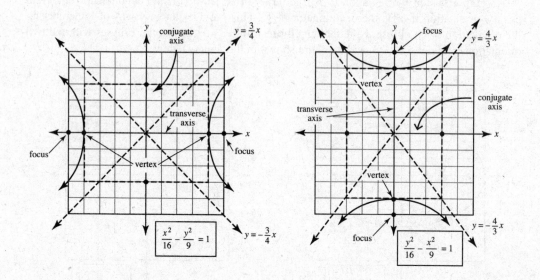

Conics can be translated as described in Chapter 12. Translations in the amounts of h and k, in the horizontal and vertical directions, are achieved by replacing x with $x - h$ and by replacing y with $y - k$ in the equation of a conic. For example, the ellipse $\dfrac{x^2}{9} + \dfrac{y^2}{16} = 1$ can be translated so that its center is $(2,-3)$ instead of $(0,0)$ by writing the equation $\dfrac{(x-2)^2}{9} + \dfrac{(y+3)^2}{16} = 1$. Each point on the ellipse defined by this equation is 2 units to the right and 3 units down from the corresponding point on $\dfrac{x^2}{9} + \dfrac{y^2}{16} = 1$. Examples (D), (E), and (F) illustrate this idea.

(A) **What is the length of the major axis in the ellipse $\dfrac{x^2}{9} + \dfrac{y^2}{4} = 1$?**

In this problem, $a^2 = 9$ is the larger of a^2 and b^2, and $a = 3$. The length of the major axis is $2a = 6$.

(B) **What are the asymptotes in the hyperbola $\dfrac{y^2}{16} - \dfrac{x^2}{25} = 1$?**

Since the y^2 term is positive, the hyperbola opens up and down, and the equations of the asymptotes are $y = \pm\dfrac{4}{5}x$.

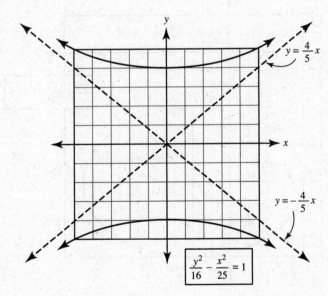

(C) What is the equation of the directrix in the parabola $x^2 = 12y$?

Since the squared term involves x and $4p = 12$ is positive, $p = 3$, and the parabola opens up. Therefore, the directrix is the horizontal line $y = -3$.

(D) What are the endpoints of the minor axis of the ellipse $\dfrac{(x-2)^2}{36} + \dfrac{(y+7)^2}{16} = 1$?

This ellipse has been translated from the origin two units to the right and seven units down, so its center is the point $(2,-7)$. Since $b^2 = 16$ is the smaller of a^2 and b^2, the ellipse is elongated horizontally, and the minor axis is vertical. The endpoints of the minor axis are $\sqrt{16} = 4$ units above and below the center. Therefore, the endpoints of the minor axis have $y = -7 + 4 = -3$ and $y = -7 - 4 = -11$, and the endpoints are $(2,-3)$ and $(2,-11)$.

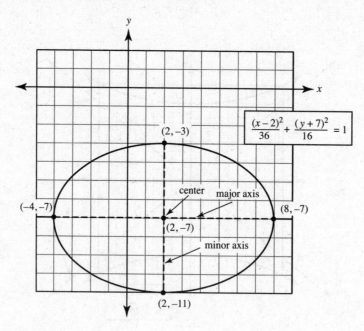

(E) What is the standard equation of the parabola with vertex (–3,5) and focus (–3,7)?

The center of this parabola is (–3,5). This means that the center of the parabola has been translated left 3 and up 5. Since the focus has the same x-coordinate as the vertex, the parabola is up or down. The focus is 2 units above the center, so $p = 2$ (which means that the parabola opens up). Therefore, the equation of the parabola is $(x + 3)^2 = 8(y - 5)$.

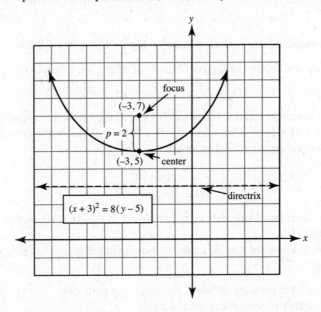

(F) What are the coordinates of the vertices of the hyperbola $\dfrac{(x-1)^2}{9} - \dfrac{(y+2)^2}{4} = 1$ **?**

The center of this hyperbola is (1,–2). The transverse axis is horizontal because the term involving x is positive. The value of $a = \sqrt{9} = 3$, and the vertices are 3 units to the right and left of the center, at (4,–2) and (–2,–2).

Conic sections can also be named by general equations of the form:

$$Ax^2 + By^2 + Cx + Dy + E = 0$$

These equations can be transformed into standard form equations by completing the square in both x and y. The following examples show how this is accomplished.

(G) Describe the conic whose equation is $x^2 + y^2 - 8x + 6y - 11 = 0$.

Complete the square in x by adding $(-4)^2 = 16$ to both sides of the equation, and complete the square in y by adding $3^2 = 9$ to both sides. After also adding 11 to both sides, the transformed equation is $(x - 4)^2 + (y + 3)^2 = 36$. This is the standard equation of a circle with center $(4, -3)$ and radius 6.

(H) Describe the conic whose equation is $9x^2 + 16y^2 - 18x + 96y + 9 = 0$.

Rearrange the terms, and factor out the coefficients of x^2 and y^2, respectively:

$$9(x^2 - 2x) + 16(y^2 + 6y) + 9 = 0$$

Then complete the square in x and y, being careful to add the constant terms to both sides of the equation: $9(x^2 - 2x + 1) + 16(y^2 + 6y + 9) + 9 - 9 - 144 = 0$. Thus, $9(x - 1)^2 + 16(y + 3)^2 = 144$. Divide both sides of this equation by 144 to get:

$$\frac{(x-1)^2}{16} + \frac{(y+3)^2}{9} = 1$$

This is an ellipse of horizontal orientation, with a center at $(1, -3)$, a major axis of length 8, and a minor axis of length 6.

If you only want to know the type of conic by looking at the general equation, there are some obvious guidelines:

- If A or B is zero, but both are not zero, the conic is a parabola.
- If $A = B$ and neither is zero, the conic is a circle.
- If A and B have opposite signs, the conic is a hyperbola.
- If $A \neq B$ and neither is zero, and if A and B have the same sign, the conic is an ellipse.

EXERCISES

1. Which of the following is a focus of $\dfrac{(x-2)^2}{4} + \dfrac{(y+1)^2}{5} = 1$?

   (A) $(1,-1)$
   (B) $(2,-1)$
   (C) $(3,-1)$
   (D) $(2,-2)$
   (E) $(-2,1)$

2. Which of the following is an asymptote of $3x^2 - 4y^2 - 12 = 0$?

   (A) $y = \dfrac{4}{3}x$

   (B) $y = -\dfrac{2}{\sqrt{3}}x$

   (C) $y = -\dfrac{3}{4}x$

   (D) $y = \dfrac{\sqrt{3}}{2}x$

   (E) $y = \dfrac{2}{3}x$

3. The standard equation of a parabola with focus $(2,-3)$ and directrix $x = 6$ is

   (A) $(y+3)^2 = 8(x-2)$
   (B) $(y+3)^2 = -8(x-4)$
   (C) $(x-2)^2 = 8(y+3)$
   (D) $(x+2)^2 = -8(y-3)$
   (E) $(x+4)^2 = -8(y-3)$

4. The standard equation of an ellipse with vertices $(-5,2)$ and $(3,2)$ and a minor axis of length 6 is

   (A) $\dfrac{(x+1)^2}{16} + \dfrac{(y-2)^2}{9} = 1$

   (B) $\dfrac{(x-1)^2}{9} + \dfrac{(y+2)^2}{16} = 1$

   (C) $\dfrac{(x+1)^2}{9} + \dfrac{(y-2)^2}{16} = 1$

   (D) $\dfrac{(x-1)^2}{16} + \dfrac{(y-2)^2}{9} = 1$

   (E) $\dfrac{(x-1)^2}{7} + \dfrac{(y+2)^2}{16} = 1$

5. Which of the following is a vertex of $16x^2 - y^2 - 32x - 6y - 57 = 0$?

   (A) $(1,-1)$
   (B) $(1,3)$
   (C) $(1,5)$
   (D) $(-1,-3)$
   (E) $(-1,3)$

6. The graph of $x^2 = (2y + 3)^2$ is

   (A) an ellipse
   (B) a parabola
   (C) a hyperbola
   (D) a circle
   (E) intersecting lines

Answers Explained

1. **(D)** This is the standard equation of an ellipse with center $(2,-1)$, $a^2 = 5$, $b^2 = 4$, and y-orientation. Since $c^2 = a^2 - b^2 = 1$, the foci are 1 unit above and below the center.

2. **(D)** Add 12 to both sides and divide both sides by 12 to put the equation in standard form:

$$\frac{x^2}{4} - \frac{y^2}{3} = 1$$

   This hyperbola has x-orientation, with $a^2 = 4$ and $b^2 = 3$. Its asymptotes are $y = \pm \dfrac{\sqrt{3}}{2} x$.

3. **(B)** The directrix is a vertical line 4 units to the right of the focus. Therefore, the parabola has an x-orientation (the y-term is square). The vertex is 2 units right of the focus at $(4,-3)$ so $p = -2$.

4. **(A)** Since the vertices have the same y-coordinate, the major axis is horizontal, has length 8, and $a^2 = 16$. Therefore, the center of the ellipse is $(-1,2)$. Since the minor axis has length 6, $b^2 = 9$.

5. **(D)** Complete the square in both x and y to get the standard equation:

$$\frac{(x-1)^2}{4} - \frac{(y+3)^2}{64} = 1$$

The transverse axis has length 4, so the vertices of the hyperbola are 2 units to the left and right of the center $(1,-3)$.

6. **(E)** Expand the right side of the equation and bring all but the constant term to the left side. This yields $x^2 - 4(y^2 + 3y) = 9$. Complete the square in y to get:

$$x^2 - 4\left(y^2 + 3y + \frac{9}{4}\right) = 9 - 9 = 0$$

Then $x^2 = (2y + 3)^2$ or $x = \pm(2y + 3)$. These are intersecting lines.

> **Reminder:** The link to the two online practice tests can be found on the inside front cover of this book. Both online tests can be accessed on most mobile devices, including tablets and smartphones.

Polar Coordinates

14

→ **DEFINITIONS**
→ **RELATIONSHIPS TO RECTANGULAR COORDINATES**

Although the most common way to represent a point in a plane is in terms of its distances from two perpendicular axes, there are several other ways. One such way is in terms of the distance of the point from the origin and the angle between the positive *x*-axis and the ray emanating from the origin going through the point.

In the figure above, the regular rectangular coordinates of *P* are (*x,y*) and the **polar coordinates** are (*r,θ*). If *r* > 0, it is measured along the ray of the terminal side of *θ*. If *r* < 0, it is measured in the direction of the opposite ray.

Since $\sin\theta = \dfrac{y}{r}$ and $\cos\theta = \dfrac{x}{r}$, there is an easy relationship between rectangular and polar coordinates:

$$x = r\cos\theta$$
$$y = r\sin\theta$$
$$x^2 + y^2 = r^2, \text{ using the Pythagorean theorem}$$

Unlike the case involving rectangular coordinates, each point in the plane can be named by an infinite number of polar coordinates.

➡ **Examples** _____

(A) (2,30°), (2,390°), (2,–330°), (–2,210°), (–2,–150°) all name the same point.

In general, a point in the plane represented by (*r,θ*) can also be represented by (*r,θ* + 2π*n*) or (–*r,θ* + (2*n* – 1)π), where *n* is an integer.

(B) Express point P, whose rectangular coordinates are $\left(3, 3\sqrt{3}\right)$, in terms of polar coordinates.

$$r^2 = x^2 + y^2 = 9 + 27 = 36$$
$$r = 6$$
$$r\cos\theta = x$$
$$\cos\theta = \frac{3}{6} = \frac{1}{2}$$

Therefore, $\theta = 60°$ and $(6, 60°)$ are the polar coordinates of P.

(C) Describe the graph of $r = 2$.

$$r^2 = x^2 + y^2$$
$$r = 2$$

Therefore, $x^2 + y^2 = 4$, which is the equation of a circle whose center is at the origin and whose radius is 2.

(D) Describe the graph of $r = \dfrac{1}{\sin\theta}$.

Since $r\sin\theta = y$ and $r = \dfrac{1}{\sin\theta}$, $y = 1$. The domain of $r = \dfrac{1}{\sin\theta}$ is all real θ that are not multiples of π.

Thus, $r = \dfrac{1}{\sin\theta}$ is the equation of a horizontal line with holes in it at multiples of π one unit above the x-axis.

Both $r = 2$ and $r = \dfrac{1}{\sin\theta}$ are examples of polar functions. Such functions can be graphed on a graphing calculator by using POLAR mode. Enter 2 into r_1 and graph using ZOOM 6, the standard window.

Although the graph is accurate for the scale given, the shape looks like that of an ellipse, not a circle. This is because the standard screen has a scale on the *x*-axis that is larger than that on the *y*-axis, resulting in a distorted graph. The graph below is obtained by using the ZOOM ZSquare command.

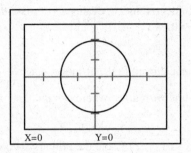

EXERCISES

1. A point has the polar coordinates (2,60°). The same point can be represented by

   (A) (−2,240°)
   (B) (2,240°)
   (C) (−2,60°)
   (D) (2,−60°)
   (E) (2,−240°)

2. The polar coordinates of a point P are (2,200°). The rectangular coordinates of P are

   (A) (−1.88,−0.68)
   (B) (−0.68,−1.88)
   (C) (−0.34,−0.94)
   (D) (−0.94,−0.34)
   (E) (−0.47,−0.17)

3. Describe the graph of $r = \dfrac{3}{\cos\theta}$.

   (A) a parabola
   (B) an ellipse
   (C) a circle
   (D) a vertical line
   (E) the *x*-axis

4. Point P has rectangular coordinates (5,12). One pair of polar coordinates for this point is

   (A) (−13,−292.6°)
   (B) (−13,−67.4°)
   (C) (−13,247.4°)
   (D) (13,−67.4°)
   (E) (13,247.4°)

5. Describe the graph of $r = 7\sin\theta$.

   (A) a parabola
   (B) a horizontal line
   (C) a circle
   (D) the *y*-axis
   (E) a hyperbola

Answers Explained

1. **(A)** The angle must either be coterminal with $(60 \pm 360n)$ with $r = 2$ or $(60 \pm 180)°$ with $r = -2$. Choice A is the only answer choice that meets these criteria.

* 2. **(A)** With your calculator in degree mode, evaluate $x = r \cos \theta = 2 \cos 200 \approx -1.88$ and $y = r \sin 200 = 2 \sin 200 \approx -0.68$.

* 3. **(D)** With your graphing calculator in POLAR mode, enter $\dfrac{3}{\cos \theta}$ as r_1, and observe that the graph is a vertical line with holes where $\cos \theta = 0$.

4. **(C)** The r-coordinate is $\sqrt{5^2 + 12^2} = 13$, and the primary θ-coordinate is $\tan^{-1}\left(\dfrac{12}{5}\right) = 67.4$, but $(13, 67.4°)$ is not an answer choice. Other correct answer choices with $r = 13$ have $\theta = 67.4° \pm 360°$. None of the answer choices with $r = 13$ have an acceptable value of θ. If $r = -13$, then a correct answer choice must have $\theta = 67.4° \pm 180°$: $67.4° + 180° = 247.4°$.

5. **(C)** Enter the equation $r = 7 \sin \theta$ into your graphing calculator, and key in ZOOM 6 to get the graph on a standard screen. The graph looks like an ellipse (not an answer choice), but because the standard screen is elongated in the horizontal direction, this graph is distorted. Key in ZOOM 5 to see that the correct answer is that the graph looks like a circle.

Three-Dimensional Geometry

15

→ **FORMULAS FOR VOLUMES AND SURFACE AREAS OF SOLIDS**
→ **RECTANGULAR COORDINATES**
→ **DISTANCE AND MIDPOINT**

The Level 2 Mathematics Subject Test may have problems that involve five types of solids: prisms, pyramids, cylinders, cones, and spheres. Rectangular solids and cubes are special types of prisms. In a rectangular solid, the bases are rectangles, so any pair of opposite sides can be bases. In a cube, the bases are squares, and the altitude is equal in length to the sides of the base.

On the Level 2 Mathematics Subject Test, all figures are assumed to be right solids. Formulas for the volumes of regular pyramids, circular cones, and spheres and the formula for the surface area of a sphere are given in the test book. For convenience, formulas for the volumes and lateral areas of all of these figures are summarized below.

Solid	Volume	Surface Area
Prism	$V = Bh$	$SA = Ph + 2B$
Rectangular solid	$V = lwh$	$SA = 2lw + 2lh + 2wh$
Cube	$V = s^3$	$SA = 6s^2$
Pyramid	$V = \dfrac{1}{3}Bh$	$SA = \dfrac{1}{2}PL + B$
Cylinder	$V = \pi r^2 h$	$SA = 2\pi rh + 2\pi r^2$
Cone	$V = \dfrac{1}{3}\pi r^2 h$	$SA = \pi rL + \pi r^2$
Sphere	$V = \dfrac{4}{3}\pi r^3$	$SA = 4\pi r^2$

The variables in these formulas are defined as follows:

V = volume h = altitude
SA = surface area l = length
B = base area w = width
P = base perimeter r = radius
L = slant height s = side length

Using the notation above, the formula for the length of a diagonal of a rectangular solid is $D = \sqrt{l^2 + w^2 + h^2}$, which is derived from two applications of the Pythagorean theorem.

Examples

(A) **The length, width, and height of a right prism are 9, 4, and 2, respectively. What is the length of the longest segment whose endpoints are vertices of the prism?**

The longest such segment is a diagonal of the prism. The length of this diagonal is:

$$l = \sqrt{9^2 + 4^2 + 2^2} = \sqrt{101}$$

One type of problem that might be found on the Level 2 Mathematics Subject Test is to find the volume of the space between two solids, one inscribed in the other. This is solved by finding the difference of volumes. In this type of problem, you are given only some of the necessary dimensions. You must use these dimensions to find the others.

(B) **A sphere with a diameter of 10 meters is inscribed in a cube. What is the volume of the space between the sphere and the cube?**

You need to subtract the volume of the sphere from the volume of the cube. The radius of the sphere is 5 and the volume is $\frac{4}{3}\pi(5)^3 \approx 523.6$ m^3. Since the sphere is inscribed in the cube, the side length of the cube is 10 m and its volume is 1,000 m^3. The volume of the space between them is 476.4 m^3.

You might be asked to find a dimension of a figure if its volume and surface area are equal.

(C) **A cylinder has a volume of 500 in^3. What is the radius of this cylinder if its altitude equals the diameter of its base?**

The formula for the volume of a cylinder is $V = \pi r^2 h$. If $h = 2r$, this formula becomes $V = 2\pi r^3$. Substitute 500 for V and solve for r to get $r = 4.3$ in.

A third type of problem involves percent changes in the dimensions of a figure.

(D) **If the volume of a right circular cone is reduced by 15% by reducing its height by 5%, by what percent must the radius of the base be reduced?**

If the volume of the cone is reduced by 15%, then 85% of its original volume remains. If its height is reduced by 5%, 95% of its original height remains. Use the formula for the volume of a cone, and let p be the proportion of the radius that remains:

$$0.85\left(\frac{1}{3}\pi r^2 h\right) = \frac{1}{3}\pi (pr)^2(0.95h)$$

This equation simplifies to $0.85 = 0.95p^2$, so $p \approx 0.946$. Therefore, the radius must be reduced by 5.4%.

A solid figure can be obtained by rotating a plane figure about some line in the plane that does not intersect the figure.

The coordinate plane can be extended by adding a third axis, the z-axis, which is perpendicular to the other two. Picture the corner of a room. The corner itself is the origin. The edges between the walls and the floor are the x- and y-axes. The edge between the two walls is the z-axis. The first octant of this three-dimensional coordinate system and the point (1,2,3) are illustrated below.

A point that has zero as any coordinate must lie on the plane formed by the other two axes. If two coordinates of a point are zero, then the point lies on the nonzero axis. The three-dimensional Pythagorean theorem yields a formula for the distance between two points (x_1, y_1, z_1) and (x_2, y_2, z_2) in space:

$$d = \sqrt{\left(x_2 - x_1\right)^2 + \left(y_2 - y_1\right)^2 + \left(z_2 - z_1\right)^2}$$

A three-dimensional coordinate system can be used to graph the variable z as a function of the two variables x and y, but such graphs are beyond the scope of the Level 2 Mathematics Subject Test.

➡ Examples

(E) The distance between two points in space A(2,y,–3) and B(1,–1,4) is 9. Find the possible values of y.

Use the formula for the distance between two points and set this equal to 9:

$$\sqrt{(2-1)^2 + (y+1)^2 + (-3-4)^2} = 9$$

Square both sides and simplify to get $(y + 1)^2 = 31$. Therefore, $y \approx 4.6$ or $y \approx -6.6$.

A sphere is the set of points in space that are equidistant from a given point. If the given point is (a,b,c) and the given distance is r, an equation of the sphere is $(x - a)^2 + (y - b)^2 + (z - c)^2 = r^2$.

(F) Describe the graph of the set of points (x,y,z) where $(x – 6)^2 + (y + 3)^2 + (z – 2)^2 = 36$.

This equation describes the set of points whose distance from (6,–3,2) is 6. This is a sphere of radius 6 with the center at (6,–3,2).

If only two of the variables appear in an equation, the equation describes a planar figure. The third variable spans the entire number line. The resulting three-dimensional figure is a solid that extends indefinitely in both directions parallel to the axis of the variable that is not in the equation, with cross sections congruent to the planar figure.

(G) Describe the graph of the set of points (x,y,z) where $x^2 + z^2 = 1$.

Since y is not in the equation, it can take any value. When restricted to the xz plane, the equation is that of a circle with radius 1 and the center at the origin. Therefore in xyz space, the equation represents a cylindrical shape, centered at the origin, that extends indefinitely in both directions along the y-axis.

A solid figure can also be obtained by rotating a plane figure about some line in the plane that does not intersect the figure.

(H) If the segment of the line $y = -2x + 2$ that lies in quadrant I is rotated about the y-axis, a cone is formed. What is the volume of the cone?

As shown in the figure below on the left, the part of the segment that lies in the first quadrant and the axes form a triangle with vertices at $(0,0)$, $(1,0)$, and $(0,2)$. Rotating this triangle about the y-axis generates the cone shown in the figure below on the right.

The radius of the base is 1, and the height is 2. Therefore, the volume is:

$$V = \frac{1}{3}\pi(1)^2(2) = \frac{2}{3}\pi$$

1. The figure below shows a right circular cylinder inscribed in a cube with edge of length x. What is the ratio of the volume of the cylinder to the volume of the cube?

x

(A) $\dfrac{2}{3}$

(B) $\dfrac{3}{4}$

(C) $\dfrac{\pi}{4}$

(D) $\dfrac{\pi}{2}$

(E) $\dfrac{4}{5}$

2. The volume of a right circular cylinder is the same numerical value as its total surface area. Find the smallest integral value for the radius of the cylinder.

(A) 1
(B) 2
(C) 3
(D) 4
(E) This value cannot be determined.

3. The length, width, and height of a rectangular solid are 5 cm, 3 cm, and 7 cm, respectively. What is the length of the longest segment whose endpoints are vertices of the rectangular solid?

(A) 5.8 cm
(B) 7.6 cm
(C) 8.6 cm
(D) 9.1 cm
(E) 15 cm

4. The distance between two points in space, $P(x,-1,-1)$ and $Q(3,-3,1)$, is 3. Find the possible values of x.

(A) 1 or 2
(B) 2 or 3
(C) -2 or -3
(D) 2 or 4
(E) -2 or -4

5. The point (–4,0,7) lies on the

   (A) *y*-axis
   (B) *xy* plane
   (C) *yz* plane
   (D) *xz* plane
   (E) *z*-axis

6. The region in the first quadrant bounded by the line $3x + 2y = 7$ and the coordinate axes is rotated about the *x*-axis. Approximately, what is the volume of the resulting solid?

   (A) 8 units³
   (B) 20 units³
   (C) 30 units³
   (D) 90 units³
   (E) 120 units³

Answers Explained

1. **(C)** The volume of the cube is x^3. The radius of the cylinder is $\dfrac{x}{2}$, and its height is x.

   Substitute these into the formula for the volume of a cylinder:

   $$V = \pi r^2 h$$

   $$= \pi \left(\frac{x}{2}\right)^2 x$$

   $$= \frac{\pi}{4} x^3$$

2. **(C)** $V = \pi r^2 h$, and the total area $= 2\pi r^2 + 2\pi rh$. Setting these two equal yields $rh = 2r + 2h$. Therefore, $h = \dfrac{2r}{r-2}$. Since h must be positive, the smallest integer value of r is 3.

* 3. **(D)** The length of the longest segment is $\sqrt{5^2 + 3^2 + 7^2} \approx 9.1$.

4. **(D)** The square of the distance between P and Q is 9, so:

   $$(x - 3)^2 + (-1 - (-3))^2 + (-1 - 1)^2 = 9, \text{ or } (x - 3)^2 = 1$$

   Therefore, $x - 3 = \pm 1$, so $x = 2$ or 4.

5. **(D)** Since the *y*-coordinate is zero, the point must lie in the *xz* plane.

* 6. **(C)** The line $3x + 2y = 7$ has *x*-intercept $\dfrac{7}{3}$ and *y*-intercept $\dfrac{7}{2}$. The part of this line that lies in the first quadrant forms a triangle with the coordinate axes. Rotating this triangle about the *x*-axis produces a cone with radius $\dfrac{7}{2}$ and height $\dfrac{7}{3}$. The volume of this cone is $\dfrac{1}{3}\pi\left(\dfrac{7}{2}\right)^2\left(\dfrac{7}{3}\right) \approx 30$.

Counting

16

→ **VENN DIAGRAMS**
→ **GENERAL MULTIPLICATION PRINCIPLE**
→ **FACTORIAL**
→ **PERMUTATIONS**
→ **COMBINATIONS**

Counting problems usually begin with the phrase "How many..." or the phrase "In how many ways..." Illustrating counting techniques by example is best.

➡ **Examples**

(A) A certain sports club has 50 members. Of these, 35 golf, 30 play soccer, and 18 do both. How many club members do neither?

Add 35 and 30, then subtract the 18 that was counted twice. This makes 47 who golf, play soccer, or do both. Therefore, only 3 (50 – 47) do neither.

(B) Among the seniors at a small high school, 80 take math, 41 take Spanish, and 54 take physics. Ten seniors take math and Spanish; 19 take math and physics; and 12 take physics and Spanish. Seven seniors take all three. How many seniors take math but not Spanish or physics?

A Venn diagram will help you sort out this complicated-sounding problem.

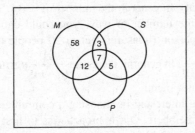

Start with the 7 who take all three courses. Since 10 take math and Spanish, this leaves 3 who take math and Spanish but not physics. Use similar reasoning to see that 5 take physics and Spanish but not math, and 12 take math and physics but not Spanish.

Finally, using the totals for how many students take each course, we can conclude that 58 (80 – 3 – 7 – 12) take math but not physics or Spanish.

Many other counting problems use the multiplication principle.

(C) Suppose you have 5 shirts, 4 pairs of pants, and 9 ties. How many outfits can be made that consist of a shirt, a pair of pants, and a tie?

For each of the 5 shirts, you can wear 4 pairs of pants, so there are 5 · 4 = 20 shirt-pants combinations. For each of these 20 shirt-pants combinations, there are 9 ties, so there are 20 · 9 = 180 shirt-pants-tie combinations.

(D) Six very good friends decide they will have lunch together every day. In how many different ways can they line up in the cafeteria line?

Any one of the 6 could be first in line. For each person who is first, there are 5 who could be second. This means there are 30 (6 · 5) ways of choosing the first two people. For each of these 30 ways, there are 4 ways of choosing the third person. This makes 120 (30 · 4) ways of choosing the first 3 people. Continuing in this fashion, there are 6 · 5 · 4 · 3 · 2 · 1 = 720 ways these 6 friends can stand in the cafeteria line. (This means that if they all have perfect attendance for 4 years of high school, they could stand in line in a different order every day, because 720 = 4 · 180.)

(E) The math team at East High has 20 members. They want to choose a president, a vice president, and a treasurer. In how many ways can this be done?

Any one of the 20 members could be president. For each choice for president, there are 19 members who could be vice president. For each of these 380 (20 · 19) ways of choosing a president and a vice president, there are 18 choices for treasurer. Therefore, there are 20 · 19 · 18 = 6,840 ways of choosing these three club officers.

(F) The student council at West High has 20 members. They want to select a committee of 3 to work with the school administration on policy matters that affect students directly. How many committees of 3 students are possible?

This problem is similar to Example (E), so we will start with the fact that if they were electing 3 officers, the student council would be able to do this in 6,840 ways. However, it does not matter whether member A is president, B is vice president, and C is treasurer or some other arrangement, as long as all 3 are on the committee. Therefore, we can divide 6,840 by the number of ways the 3 students selected could be president, vice president, and treasurer. This latter number is 3 · 2 · 1 = 6, so there are 1,140 (6,840 ÷ 6) committees of 3.

Counting problems like the ones in Examples (D), (E), and (F) occur frequently enough that they have special designations.

Example (D) asked for the number of ways 6 friends could stand in line. By using the multiplication principle, we found that there were 6 · 5 · 4 · 3 · 2 · 1 ways. A special notation for this product is 6! (6 **factorial**). In general, the number of ways "n" objects can be ordered is $n!$.

Example (E) asked for the number of ways you could choose a first (president), second (vice president), and third person (treasurer) out of 20 people ($r = 3$, $n = 20$). The answer is $20 \cdot 19 \cdot 18$, or $\dfrac{20!}{17!} = \dfrac{20!}{(20-3)!}$. In general, there are $\dfrac{n!}{(n-r)!}$ **permutations** of r objects of n. This appears as $_nP_r$ in the calculator menu.

In Example (F), we were interested in choosing a committee of 3 where there was no distinction among committee members. Our approach was to first compute the number of ways of choosing officers and then divide out the number of ways the three officers could hold the different offices. This led to the computation $\dfrac{20!}{17!3!} = \dfrac{20!}{(20-3)!3!}$. In general, the number of ways of choosing r of n objects is $\dfrac{n!}{(n-r)!r!}$. This quantity appears on the calculator menu as $_nC_r$. However, there is a special notation for **combinations**: $_nC_r = \dbinom{n}{r}$ = the number of ways r objects can be chosen from n.

Calculator commands for all three of these functions are in the MATH/PRB menu.

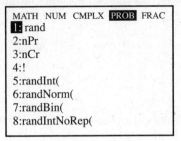

These 3 commands can also be found on scientific calculators.

EXERCISES

1. There are 50 people in a room. Twenty-eight are male, and 32 are under the age of 30. Twelve are males under the age of 30. How many women over the age of 30 are in the group?

   (A) 2
   (B) 3
   (C) 4
   (D) 5
   (E) 6

2. A student has 8 shirts, 5 pairs of pants, and 3 pairs of shoes. How many outfits that consist of a shirt, a pair of pants, and a pair of shoes can this student form?

   (A) 16
   (B) 120
   (C) 560
   (D) 4,096
   (E) 2,903,400

3. Plain candies come in six colors: brown, green, orange, red, blue, and yellow. Assume there are at least 3 of each color. If you pick three candies from a bag, how many color possibilities are there?

   (A) 18
   (B) 20
   (C) 120
   (D) 216
   (E) 729

4. A code consists of two letters of the alphabet followed by 5 digits. How many such codes are possible?

   (A) 7
   (B) 10
   (C) 128
   (D) 20,000
   (E) 67,600,000

5. A salad bar has 7 ingredients, excluding the dressing. How many different salads are possible where two salads are different if they don't include identical ingredients?

(A) 7
(B) 14
(C) 128
(D) 5,040
(E) 823,543

6. How many 3-person committees can be selected from a fraternity with 25 members?

(A) 15,625
(B) 13,800
(C) 2,300
(D) 75
(E) 8

7. A basketball team has 5 centers, 9 guards, and 13 forwards. Of these, 1 center, 2 guards, and 2 forwards start a game. How many possible starting teams can a coach put on the floor?

(A) 56,160
(B) 14,040
(C) 585
(D) 197
(E) 27

8. Five boys and 6 girls would like to serve on the homecoming court, which will consist of 2 boys and 2 girls. How many different homecoming courts are possible?

(A) 30
(B) 61
(C) 150
(D) 900
(E) 2,048

9. In a plane there are 8 points, no three of which are collinear. How many lines do the points determine?

(A) 7
(B) 16
(C) 28
(D) 36
(E) 64

10. If $\binom{6}{x} = \binom{4}{x}$, then $x =$

(A) 0
(B) 1
(C) 4
(D) 5
(E) 10

Answers Explained

1. **(A)** A Venn diagram will help you solve this problem.

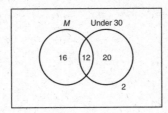

The two circles represent males and people who are at most 30 years of age, respectively. The part of the rectangle outside both circles represents people who are in neither category, i.e., females over the age of 30. First fill in the 12 males who are less than or equal to 30 years of age in the intersection of the circles. Since there are 28 males altogether, 16 are male and over 30. Since there are 32 people age 30 or less, there are 20 women that age. Add these together to get 48 people. Since there are 50 in the group, 2 must be women over 30.

2. **(B)** This is a direct application of the multiplication rule: $8 \times 5 \times 3 = 120$.

✳ 3. **(D)** There are 6 choices of color for each of the three candies selected. Therefore, there are $6 \times 6 \times 6 = 216$ color possibilities altogether.

✳ 4. **(E)** The multiplication rule applies. There are $(26)(26)(10)^5 = 67,600,000$ possible codes.

✳ 5. **(C)** You can either include or exclude each of the seven ingredients in your salad, which means there are 2 choices for each ingredient. According to the multiplication rule, there are $2^7 = 128$ ways of making these yes-no choices.

✳ 6. **(C)** This is the number of ways 3 people can be chosen from 25:

$$\binom{25}{3} = 25_nC_r3 = 2,300$$

✳ 7. **(B)** There are $\binom{5}{1} = 5$ ways of choosing the one center, $\binom{9}{2} = 36$ ways of choosing the two guards, and $\binom{13}{2} = 78$ ways of choosing the two forwards. Therefore, there are $5 \times 36 \times 78 = 14,040$ possible starting teams.

✳ 8. **(C)** There are $\binom{5}{2} = 10$ ways of choosing 2 boys out of 5 and $\binom{6}{2} = 15$ ways of choosing 2 girls out of 6. Therefore, there are $10 \times 15 = 150$ ways of choosing the homecoming court.

✳ 9. **(C)** Since no three points are collinear, every pair of points determines a distinct line. There are $\binom{8}{2} = 28$ such lines.

10. **(A)** $\binom{n}{0} = 1$ for any n.

Complex Numbers

17

→ **POWERS OF *i***
→ **ARITHMETIC OF COMPLEX NUMBERS**
→ **GRAPHING COMPLEX NUMBERS**

The square of a real number is never negative. This means that the square root of a negative number cannot be a real number. The symbol $i = \sqrt{-1}$ is called the **imaginary unit**, $i^2 = -1$. Powers of i follow a pattern:

Power of *i*	Intermediate Steps	Value
i^1	i	i
i^2	$i \cdot i = -1$	-1
i^3	$i^2 \cdot i = (-1) \cdot i = -i$	$-i$
i^4	$i^3 \cdot i = (-i) \cdot i = -i^2 = -(-1) = 1$	1
i^5	$i^4 \cdot i = 1 \cdot i = i$	i

In other words, powers of i follow a cycle of four. This means that $i^n = i^{n \bmod 4}$, where $n \bmod 4$ is the remainder when n is divided by 4. For example, $i^{58} = i^2 = -1$.

The imaginary numbers are numbers of the form bi, where b is a real number. The square root of any negative number is i times the square root of the positive of that number. Thus for example, $\sqrt{-9} = i\sqrt{9} = 3i$, $\sqrt{-12} = i\sqrt{12} = 2i\sqrt{3}$, and $\sqrt{-7} = i\sqrt{7}$.

The complex numbers are formed by "attaching" imaginary numbers to real numbers using a plus sign (+). The standard form of a complex number is $a + bi$, where a and b are real. The number a is called the real part of the complex number, and the b number is called the imaginary part. If $b = 0$, then the complex number is just a real number. If $b \neq 0$, the complex number is called imaginary. If $a = 0$, bi is called a pure imaginary number. Examples of imaginary numbers are $2 + 3i, -\dfrac{3}{5} + 4i, 6i, 0.11 + (-0.45)i$, and $\pi - i\sqrt{5}$.

When the imaginary part of a complex number is a radical, write the i to the left in order to avoid ambiguity about whether i is under the radical.

Finding sums, differences, products, quotients, and reciprocals of complex numbers can be accomplished directly on your calculator. The imaginary unit i is the 2nd decimal point on the graphing calculator. If you enter an expression with i in it, the calculator will do imaginary arithmetic in REAL mode. If the expression entered does not include i but the output is imaginary, the calculator will give you the error message NONREAL ANS. For example, if you tried to calculate $\sqrt{-3}$ in REAL mode, you would get this error message. In $a + bi$ mode, however, $\sqrt{-3}$ would calculate as $1.732\cdots i$. You should use $a + bi$ mode exclusively for the Level 2 Mathematics Subject Test. Although complex number arithmetic per se is not likely to be on the Level 2 Mathematics Subject Test, an understanding of how it is done may be. A review of the main features of complex number arithmetic is provided in the next several paragraphs.

To add or subtract complex numbers, add or subtract their real and imaginary parts.

➡ Examples

(A) Evaluate (5 – 7i) + (2 + 4i).

$$(5 - 7i) + (2 + 4i) = 7 - 3i$$

To multiply complex numbers, multiply like you would any two binomial expressions, using FOIL. Thus:

$$(a + bi)(c + di) = ac + adi + bci + bdi^2 = (ac - bd) + (ad + bc)i$$

(B) Evaluate (4 + 5i)(2 – 3i).

$$(4 + 5i)(2 - 3i) = 8 + 10i - 12i - 15i^2 = 8 - 2i + 15 = 23 - 2i$$

The difference of the first and last terms makes the real part, and the sum of the outer and inner terms makes the imaginary part.

To find the quotient of two complex numbers, multiply the denominator and the numerator by the conjugate of the denominator. Then simplify.

(C) Evaluate $\dfrac{2 - 7i}{3 + 5i}$.

$$\frac{2 - 7i}{3 + 5i} = \left(\frac{2 - 7i}{3 + 5i}\right) \cdot \left(\frac{3 - 5i}{3 - 5i}\right) = \frac{-29 - 31i}{9 + 25} = -\frac{29}{34} - \frac{31}{34}i$$

A complex number can be represented graphically as rectangular coordinates, with the x-coordinate as the real part and the y-coordinate as the imaginary part. The modulus of a complex number is its distance to the origin. The Pythagorean theorem tells us that this distance is $\sqrt{a^2 + b^2}$. The conjugate of the imaginary number $a + bi$ is $a - bi$, so the graphs of conjugates are reflections about the y-axis. Also, the product of an imaginary number and its conjugate is the square of the modulus because $(a + bi)(a - bi) = a^2 - b^2i^2 = a^2 + b^2$.

EXERCISES

1. $i^{29} =$

   (A) 1
   (B) i
   (C) $-i$
   (D) -1
   (E) none of the above

2. Evaluate $(2 + 3i)(4 - 5i)$.

   (A) $-7 - 23i$
   (B) $-7 + 2i$
   (C) $23 - 7i$
   (D) $23 + 2i$
   (E) $23 - 2i$

3. Evaluate $\dfrac{i}{2-i}$.

   (A) $-1 + \dfrac{1}{2}i$

   (B) $\dfrac{1}{5} - \dfrac{2}{5}i$

   (C) $-\dfrac{1}{5} + \dfrac{2}{5}i$

   (D) $-1 + 2i$

   (E) $-1 - 2i$

4. If $z = 8 - 2i$, $z^2 =$

   (A) $60 - 32i$
   (B) $64 + 4i$
   (C) $64 - 4i$
   (D) 60
   (E) 68

5. If z is the complex number shown in the figure below, which of the following points could be iz?

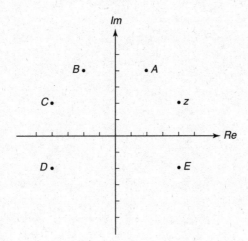

   (A) A
   (B) B
   (C) C
   (D) D
   (E) E

6. Which of the following is the modulus of $2 + i$?

   (A) $\sqrt{2}$
   (B) 2
   (C) $\sqrt{3}$
   (D) $\sqrt{5}$
   (E) 5

Answers Explained

1. **(B)** $i^{29} = i^1 = i$.

2. **(D)** If you enter imaginary numbers into the calculator, it will do imaginary arithmetic without changing mode. The imaginary unit is the 2nd decimal point. Enter the product, and read the solution, $23 + 2i$.

* 3. **(C)** Simply enter the expression into the graphing calculator.

* 4. **(A)** Simply enter the expression into the graphing calculator.

5. **(B)** $z = 4 + 2i$, so $iz = -2 + 4i$, which is point B.

6. **(D)** The real and imaginary parts are 2 and 1, respectively, so the modulus is $\sqrt{2^2 + 1^2} = \sqrt{5}$.

Matrices

18

→ **ADDITION AND SUBTRACTION**
→ **SCALAR MULTIPLICATION**
→ **MATRIX MULTIPLICATION**
→ **DETERMINANTS**
→ **SYSTEMS OF EQUATIONS**

A matrix is a rectangular array of numbers. The size of a matrix is r by c, where r is the number of rows and c is the number of columns. The numbers in a matrix are called entries, and the entry in the ith row and jth column is named x_{ij}. Two matrices are equal if they are the same size and their corresponding entries are equal.

➥ Examples

(A) **Evaluate x and y if** $\begin{bmatrix} 5 & -3 \\ x & 4 \end{bmatrix} = \begin{bmatrix} y-3 & -3 \\ 2x+2 & 4 \end{bmatrix}$.

These two matrices are equal if $y - 3 = 5$ and $x = 2x + 2$. Therefore, $y = 8$ and $x = -2$.

If $r = 1$, the matrix is called a row matrix. If $c = 1$, the matrix is called a column matrix. If $r = c$, the matrix is called a square matrix. The numbers from the upper left corner to the bottom right corner of a square matrix form the main diagonal.

Scalar multiplication takes place when each number in a matrix is multiplied by a constant. If two matrices are the same size, they can be added or subtracted by adding or subtracting corresponding entries.

(B) **Simplify:** $3\begin{bmatrix} -2 & 3 \\ 1 & 5 \\ -4 & 3 \end{bmatrix} - 2\begin{bmatrix} 3 & -1 \\ 2 & 1 \\ -4 & 6 \end{bmatrix}$.

$$3\begin{bmatrix} -2 & 3 \\ 1 & 5 \\ -4 & 3 \end{bmatrix} - 2\begin{bmatrix} 3 & -1 \\ 2 & 1 \\ -4 & 6 \end{bmatrix} = \begin{bmatrix} -6 & 9 \\ 3 & 15 \\ -12 & 9 \end{bmatrix} - \begin{bmatrix} 6 & -2 \\ 4 & 2 \\ -8 & 12 \end{bmatrix} = \begin{bmatrix} -12 & 11 \\ -1 & 13 \\ -4 & -3 \end{bmatrix}$$

(C) Solve the matrix equation: $2x + \begin{bmatrix} -3 & 2 & 6 \\ 5 & -1 & 0 \\ 3 & -6 & -2 \end{bmatrix} = \begin{bmatrix} 5 & -8 & -4 \\ 1 & 3 & 10 \\ -5 & 8 & 0 \end{bmatrix}$.

$$2x = \begin{bmatrix} 5 & -8 & -4 \\ 1 & 3 & 10 \\ -5 & 8 & 0 \end{bmatrix} - \begin{bmatrix} -3 & 2 & 6 \\ 5 & -1 & 0 \\ 3 & -6 & -2 \end{bmatrix} = \begin{bmatrix} 8 & -10 & -10 \\ -4 & 4 & 10 \\ -8 & 14 & 2 \end{bmatrix}$$

$$x = \begin{bmatrix} 4 & -5 & -5 \\ -2 & 2 & 5 \\ -4 & 7 & 1 \end{bmatrix}$$

Matrix multiplication takes place when two matrices, A and B, are multiplied to form a new matrix, AB. Matrix multiplication is possible only under certain conditions. Suppose A is r_1 by c_1 and B is r_2 by c_2. If $c_1 = r_2$, then AB is defined and has size r_1 by c_2. The entry x_{ij} of AB is the ith row of A times the jth column of B. If A and B are square matrices, BA is also defined but not generally equal to AB.

➡ **Example** _____

(D) Evaluate $AB = \begin{bmatrix} 3 & -1 \\ 3 & 5 \\ -2 & 1 \end{bmatrix} \begin{bmatrix} 1 & -2 \\ 5 & -3 \end{bmatrix}$. A is 3 by 2, and B is 2 by 2.

The product matrix is 3 by 2, with entries x_{ij} calculated as follows:

$x_{11} = (3)(1) + (-1)(5) = -2$, the "product" of row 1 of A and column 1 of B

$x_{12} = (3)(-2) + (-1)(-3) = -3$, the "product" of row 1 of A and column 2 of B

$x_{21} = (3)(1) + (5)(5) = 28$, the "product" of row 2 of A and column 1 of B

$x_{22} = (3)(-2) + (5)(-3) = -21$, the "product" of row 2 of A and column 2 of B

$x_{31} = (-2)(1) + (1)(5) = 3$, the "product" of row 3 of A and column 1 of B

$x_{32} = (-2)(-2) + (1)(-3) = 1$, the "product" of row 3 of A and column 2 of B

Therefore, $AB = \begin{bmatrix} 3 & -1 \\ 3 & 5 \\ -2 & 1 \end{bmatrix} \begin{bmatrix} 1 & -2 \\ 5 & -3 \end{bmatrix} = \begin{bmatrix} -2 & -3 \\ 28 & -21 \\ 3 & 1 \end{bmatrix}$. Note that BA is not defined.

Matrix calculations can be done on a graphing calculator. To define a matrix, enter 2nd MATRIX, highlight and enter EDIT, enter the number of rows followed by the number of columns, and finally enter the entries. The figures below show the result of these steps for matrices A and B of Example (D).

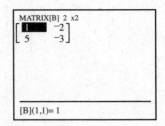

To find the product, enter 2nd MATRIX/NAMES/[A], which returns [A] to the home screen. Also enter 2nd MATRIX/NAMES/[B], which returns [B] to the home screen. Hit ENTER again to get the product.

$$
\begin{array}{l}
\text{[A][B]} \\[4pt]
\begin{bmatrix} -2 & -3 \\ 28 & -21 \\ 3 & 1 \end{bmatrix}
\end{array}
$$

Square matrices of the same size can always be multiplied. The product of square n by n matrices is a square n by n matrix. An identity matrix I_n is a square matrix consisting of 1s down the main diagonal and 0s elsewhere. The product of n by n square matrices A and I_n is A. In other words, I_n is a multiplicative identity for n by n square matrices.

➡ **Example** _____

(E) **The identity matrix** $I_2 = \begin{bmatrix} 1 & 0 \\ 0 & 1 \end{bmatrix}$. **If** $A = \begin{bmatrix} 2 & -3 \\ 5 & 1 \end{bmatrix}$, **evaluate** $AI_2 = I_2A = A$.

$$
AI_n = \begin{bmatrix} 2 & -3 \\ 5 & 1 \end{bmatrix}\begin{bmatrix} 1 & 0 \\ 0 & 1 \end{bmatrix} = \begin{bmatrix} 2 & -3 \\ 5 & 1 \end{bmatrix} \text{ and } I_nA = \begin{bmatrix} 1 & 0 \\ 0 & 1 \end{bmatrix}\begin{bmatrix} 2 & -3 \\ 5 & 1 \end{bmatrix} = \begin{bmatrix} 2 & -3 \\ 5 & 1 \end{bmatrix}
$$

In general, matrix multiplication is *not* commutative.

➡ **Example** _____

(F) **If** $A = \begin{bmatrix} -2 & 5 \\ 1 & 3 \end{bmatrix}$ **and** $B = \begin{bmatrix} 6 & 0 \\ 3 & -5 \end{bmatrix}$, **evaluate** AB **and** BA.

$$
AB = \begin{bmatrix} -2 & 5 \\ 1 & 3 \end{bmatrix}\begin{bmatrix} 6 & 0 \\ 3 & -5 \end{bmatrix} = \begin{bmatrix} 3 & -25 \\ 15 & -15 \end{bmatrix}, \text{ while } BA = \begin{bmatrix} 6 & 0 \\ 3 & -5 \end{bmatrix}\begin{bmatrix} -2 & 5 \\ 1 & 3 \end{bmatrix} = \begin{bmatrix} -12 & 30 \\ -11 & 0 \end{bmatrix}.
$$

The **determinant** of an n by n square matrix is a number. The determinant of the 2 by 2 matrix $\begin{bmatrix} a & b \\ c & d \end{bmatrix}$ is denoted by $\begin{vmatrix} a & c \\ b & d \end{vmatrix}$, which equals $ad - bc$. Determinants of larger square matrices can be evaluated by keying 2nd/MATRIX/MATH/det on the graphing calculator.

➡ **Examples** _____

(G) **Write an expression for the determinant of** $\begin{bmatrix} 2 & -1 \\ 3 & x \end{bmatrix}$.

By definition, $\begin{vmatrix} 2 & -1 \\ 3 & x \end{vmatrix} = 2x - (-3) = 2x + 3$.

(H) Solve for x: $\begin{vmatrix} x & x \\ 8 & x \end{vmatrix} = \begin{vmatrix} 7 & -2 & 1 \\ 0 & 3 & -1 \\ 5 & -4 & 2 \end{vmatrix}$.

The determinant on the left side is $x^2 - 8x$. Use the calculator to evaluate the determinant on the right as 9. This yields the quadratic equation $x^2 - 8x - 9 = 0$. This can be solved by factoring to get $x = 9$ or $x = -1$.

A square matrix is called **singular** if its determinant is zero and **nonsingular** if its determinant is not zero. **Only nonsingular matrices have inverses.** The **inverse of a nonsingular square matrix**, A, is defined as a square matrix A^{-1} of the same size such that $AA^{-1} = A^{-1}A = I$.

The inverse of a nonsingular square n by n matrix A has a multiplicative inverse, A^{-1}, where $A^{-1}A = AA^{-1} = I$. A^{-1} can be found on the graphing calculator by entering MATRIX/NAMES/A, which will return A to the home screen, followed by x^{-1} and ENTER.

(I) If $A = \begin{bmatrix} 7 & -3 \\ 1 & 4 \end{bmatrix}$ and $B = \begin{bmatrix} 5 \\ 2 \end{bmatrix}$, solve for X when $AX = B$.

Matrix multiply both sides on the left by A^{-1}: $A^{-1}AX = A^{-1}B$. This yields $IX = X = A^{-1}B = \begin{bmatrix} 26/31 \\ 9/31 \end{bmatrix}$.

The fractional form of the answer can be obtained by keying MATH/ENTER/ENTER.

An important application of matrices is writing and solving systems of equations in matrix form.

➡ **Example** _____

(J) Solve the system $\begin{aligned} x - y + 2z &= -3 \\ 2x + y - z &= 0 \\ -x + 2y - 3z &= 7 \end{aligned}$

The matrix form of this system is $AX = B$, where $A = \begin{bmatrix} 1 & -1 & 2 \\ 2 & 1 & -1 \\ -1 & 2 & -3 \end{bmatrix}$, $X = \begin{bmatrix} x \\ y \\ z \end{bmatrix}$, and

$B = \begin{bmatrix} -3 \\ 0 \\ 7 \end{bmatrix}$. A is the **coefficient matrix**, X is the **variables matrix**, and B is the **constant matrix**.

This system can be represented by the single 3×4 matrix $\begin{bmatrix} 1 & -1 & 2 & -3 \\ 2 & 1 & -1 & 0 \\ -1 & 2 & -3 & 7 \end{bmatrix}$. In general, if there were n equations and n variables, then matrix A would have n rows and $n + 1$ columns.

You can use the calculator to transform this matrix to **reduced row echelon form**. This transformation involves multiplying a row by a constant, or multiplying a row by a constant and adding the result to another row, and replacing the second row with the sum. (This is the same procedure used to solve a system of two equations in two variables.) These operations are done with the intent of transforming the first three columns into a 3×3 identity matrix. As this is done, the fourth column transforms into the solution. In the example above:

$$\begin{bmatrix} 1 & -1 & 2 & -3 \\ 2 & 1 & -1 & 0 \\ -1 & 2 & -3 & 7 \end{bmatrix} \rightarrow \begin{bmatrix} 1 & 0 & 0 & -2 \\ 0 & 1 & 0 & 7 \\ 0 & 0 & 1 & 3 \end{bmatrix}, \text{ so } x = -2, y = 7, z = 3$$

1. $\begin{bmatrix} 1 & 3 \\ -2 & 4 \end{bmatrix} + \begin{bmatrix} 11 & 5 \\ -6 & 12 \end{bmatrix} = K\begin{bmatrix} 3 & 2 \\ J & M \end{bmatrix}$. Find the value of $K + J + M$.

(A) 2
(B) 4
(C) 6
(D) 7
(E) 8

2. Evaluate x and y if $\begin{bmatrix} x & 2 \\ -3 & y \end{bmatrix} = 2\begin{bmatrix} x^2 & 1 \\ -\dfrac{3}{2} & 3y-5 \end{bmatrix}$.

(A) $x = 0; y = 2$
(B) $x = 1; y = 2$
(C) $x = -1, 1; y = \dfrac{5}{3}$

(D) $x = -\dfrac{1}{2}, \dfrac{1}{2}; y = \dfrac{5}{6}$

(E) $x = 0, \dfrac{1}{2}; y = 2$

3. Solve for X: $\begin{bmatrix} 1 & 2 & -3 \\ 2 & 1 & 3 \end{bmatrix} - X = \begin{bmatrix} 5 & 1 & 8 \\ -6 & 0 & 5 \end{bmatrix}$.

(A) $\begin{bmatrix} 4 & 1 & -11 \\ -8 & 1 & -2 \end{bmatrix}$

(B) $\begin{bmatrix} -4 & 1 & -11 \\ 8 & 1 & -2 \end{bmatrix}$

(C) $\begin{bmatrix} -5 & -2 & 24 \\ 12 & 0 & -15 \end{bmatrix}$

(D) $\begin{bmatrix} 5 & 2 & -24 \\ -12 & 0 & 15 \end{bmatrix}$

(E) $\begin{bmatrix} 6 & 3 & 5 \\ -4 & 1 & 8 \end{bmatrix}$

Use $A = \begin{bmatrix} -2 & -1 & 5 & 9 \end{bmatrix}$ and $B = \begin{bmatrix} 0 & -5 \\ 3 & -2 \\ 4 & 0 \\ -6 & 1 \end{bmatrix}$ for question 4.

4. The product $AB =$

(A) $\begin{bmatrix} 0 & 10 \\ -3 & 2 \\ 20 & 0 \\ -54 & 9 \end{bmatrix}$

(B) $\begin{bmatrix} -37 & 21 \end{bmatrix}$

(C) $\begin{bmatrix} 10 \\ -2 \\ 20 \\ -45 \end{bmatrix}$

(D) $\begin{bmatrix} 0 \\ 6 \\ 0 \\ -54 \end{bmatrix}$

(E) product is not defined

5. The first row, second column of the product $\begin{bmatrix} x & 1 \\ 2 & -3 \end{bmatrix}\begin{bmatrix} 5 & -x \\ 2 & 1 \end{bmatrix}$ is

(A) $-5x - 3$
(B) $-x - 3$
(C) $1 - x^2$
(D) $4x$
(E) $2x + 2$

6. If $A = \begin{bmatrix} -3 & 1 & 6 \\ 2 & -5 & 0 \\ 1 & -3 & 4 \end{bmatrix}$, $B = \begin{bmatrix} 4 & 7 \\ -4 & 2 \\ -1 & -5 \end{bmatrix}$, and $AX = B$, then the size of X is

(A) 3 rows, 3 columns
(B) 3 rows, 2 columns
(C) 2 rows, 2 columns
(D) 2 rows, 3 columns
(E) X cannot be determined.

7. The chart below shows the number of small and large packages of a certain brand of cereal that were bought over a three-day period. The price of a small box of this brand is $2.99, and the price of a large box is $3.99. Which of the following matrix expressions represents the income, in dollars, received from the sale of cereal each of the three days?

	Day 1	Day 2	Day 3
Large	75	82	57
Small	43	36	50

(A) $\begin{bmatrix} 75 & 82 & 57 \\ 43 & 36 & 50 \end{bmatrix} \begin{bmatrix} 2.99 & 3.99 \end{bmatrix}$

(B) $\begin{bmatrix} 75 & 43 \\ 82 & 36 \\ 57 & 50 \end{bmatrix} \begin{bmatrix} 3.99 \\ 2.99 \end{bmatrix}$

(C) $\begin{bmatrix} 75 & 82 & 57 \\ 43 & 36 & 50 \end{bmatrix} \begin{bmatrix} 2.99 \\ 3.99 \end{bmatrix}$

(D) $\begin{bmatrix} 2.99 \\ 3.99 \end{bmatrix} \begin{bmatrix} 75 & 43 \\ 82 & 36 \\ 57 & 50 \end{bmatrix}$

(E) $2.99 \begin{bmatrix} 75 & 82 & 57 \\ 43 & 36 & 50 \end{bmatrix} + 3.99 \begin{bmatrix} 75 & 82 & 57 \\ 43 & 36 & 50 \end{bmatrix}$

8. The determinant of $\begin{bmatrix} p & 3 \\ -2 & 1 \end{bmatrix}$ is
   (A) $p - 6$
   (B) $p + 6$
   (C) $3p - 2$
   (D) $3 - 2p$
   (E) $-6 - p$

9. Find all values of x for which $\begin{vmatrix} 2 & -1 & 4 \\ 3 & 0 & 5 \\ 4 & 1 & 6 \end{vmatrix} = \begin{vmatrix} x & 4 \\ 5 & x \end{vmatrix}$.

   (A) ±3.78
   (B) ±4.47
   (C) ±5.12
   (D) ±6.19
   (E) ±6.97

10. If $X \begin{bmatrix} -7 & 2 \\ 0 & -5 \end{bmatrix} = \begin{bmatrix} 2 & -3 \\ 5 & 4 \end{bmatrix}$, then $X =$

(A) $\begin{bmatrix} -\dfrac{2}{7} & -\dfrac{3}{2} \\ 0 & -\dfrac{5}{4} \end{bmatrix}$

(B) $\begin{bmatrix} -\dfrac{7}{2} & -\dfrac{2}{3} \\ 0 & -\dfrac{4}{5} \end{bmatrix}$

(C) $\begin{bmatrix} -\dfrac{2}{7} & \dfrac{17}{35} \\ -\dfrac{5}{7} & -\dfrac{38}{35} \end{bmatrix}$

(D) $\begin{bmatrix} -14 & -6 \\ 0 & -20 \end{bmatrix}$

(E) undefined

11. Find the matrix equation that represents the system $\begin{cases} 2x - 3 = 3y \\ y - 5x = 14 \end{cases}$

(A) $\begin{bmatrix} 2 & -3 \\ 1 & 5 \end{bmatrix} \begin{bmatrix} x \\ y \end{bmatrix} = \begin{bmatrix} 3 \\ 14 \end{bmatrix}$

(B) $\begin{bmatrix} 2 & -3 \\ -5 & 1 \end{bmatrix} \begin{bmatrix} x \\ y \end{bmatrix} = \begin{bmatrix} 3 \\ 14 \end{bmatrix}$

(C) $\begin{bmatrix} x \\ y \end{bmatrix} \begin{bmatrix} 2 & -3 \\ -5 & 1 \end{bmatrix} = \begin{bmatrix} 3 \\ 14 \end{bmatrix}$

(D) $\begin{bmatrix} 3 \\ 13 \end{bmatrix} \begin{bmatrix} x & y \end{bmatrix} = \begin{bmatrix} 2 & -3 \\ 5 & 1 \end{bmatrix}$

(E) This system cannot be represented as a matrix equation.

12. Find $\begin{bmatrix} x \\ y \end{bmatrix}$ if $\begin{bmatrix} 3 & 2 \\ -1 & 4 \end{bmatrix}\begin{bmatrix} x \\ y \end{bmatrix} = \begin{bmatrix} -5 \\ 4 \end{bmatrix}$.

(A) $[-2 \quad 0.5]$

(B) $\begin{bmatrix} -\dfrac{5}{6} \\ 1 \end{bmatrix}$

(C) $\begin{bmatrix} -1 & \dfrac{3}{4} \end{bmatrix}$

(D) $\begin{bmatrix} -2 \\ \dfrac{1}{2} \end{bmatrix}$

(E) $\begin{bmatrix} -5 & -\dfrac{4}{5} \end{bmatrix}$

Answers Explained

1. **(C)** First find the sum of the matrices to the left of the equal sign: $\begin{bmatrix} 12 & 8 \\ -8 & 16 \end{bmatrix}$.

   Since the first row of the matrix to the right of the equal sign is $(3 \quad 2)$, K must be 4. Since $(J \quad M)$ is the bottom row, $J = -2$ and $M = 4$. Therefore, $K + J + M = 6$.

2. **(E)** In order for these matrices to be equal, $\begin{bmatrix} x & 2 \\ -3 & y \end{bmatrix} = \begin{bmatrix} 2x^2 & 2 \\ -3 & 6y-10 \end{bmatrix}$.

   Therefore, $x = 2x^2$ and $y = 6y - 10$. Solving the first equation yields $x = 0, \dfrac{1}{2}$ and $y = 2$.

3. **(B)** To solve for X, first subtract $\begin{bmatrix} 1 & 2 & -3 \\ 2 & 1 & 3 \end{bmatrix}$ from both sides of the equation.

   Then $-X = \begin{bmatrix} 4 & -1 & 11 \\ -8 & -1 & 2 \end{bmatrix}$, so $X = \begin{bmatrix} -4 & 1 & -11 \\ 8 & 1 & -2 \end{bmatrix}$.

4. **(B)** By definition, $AB = \begin{bmatrix} (-2)(0)+(-1)(3)+(5)(4)+(9)(-6) \\ (-2)(-5)+(-1)(-2)+(5)(0)+(9)(1) \end{bmatrix} = \begin{bmatrix} -37 & 21 \end{bmatrix}$.

5. **(C)** By definition, the first row, second column of the product is $(x)(-x) + (1)(1) = -x^2 + 1$.

6. **(B)** X must have as many rows as A has columns, which is 3. X must have as many columns as B does, which is 2.

7. **(B)** Matrix multiplication is row by column. Since the answer must be a 3 by 1 matrix, the only possible answer choice is Choice B.

8. **(B)** By definition, the determinant of $\begin{bmatrix} p & 3 \\ -2 & 1 \end{bmatrix}$ is $(p)(1) - (3)(-2) = p + 6$.

9. **(B)** Enter the 3 by 3 matrix on the left side of the equation into your graphing calculator and evaluate its determinant (zero). The determinant on the right side of the equation is $x^2 - 20$. Therefore, $x = \pm\sqrt{20} \approx \pm 4.47$.

* 10. **(C)** To find X, multiply both sides of the equation by $\begin{bmatrix} -7 & 2 \\ 0 & -5 \end{bmatrix}^{-1}$ on the right. Enter

both matrices in your calculator, key the product $\begin{bmatrix} 2 & -3 \\ 5 & 4 \end{bmatrix} \begin{bmatrix} -7 & 2 \\ 0 & -5 \end{bmatrix}^{-1}$ on your

graphing calculator, and key MATH/ENTER/ENTER to convert the decimal answer to a fraction.

11. **(B)** First, write the system in standard form: $\begin{cases} 2x - 3y = 3 \\ -5x + y = 14 \end{cases}$. The matrix form of this

equation is $\begin{bmatrix} 2 & -3 \\ -5 & 1 \end{bmatrix} \begin{bmatrix} x \\ y \end{bmatrix} = \begin{bmatrix} 3 \\ 14 \end{bmatrix}$.

12. **(D)** This is the matrix form $AX = B$ of a system of equations. Multiply both sides of

the equation by A^{-1} on the left to get the solution $X = \begin{bmatrix} x \\ y \end{bmatrix} = A^{-1}B$. Enter the 2 by 2

matrix, A, and the 2 by 1 matrix, B, into your graphing calculator. Return to the home

screen and enter $A^{-1}B = \begin{bmatrix} -2 \\ 0.5 \end{bmatrix}$.

Sequences and Series

19

- → **SIGMA NOTATION**
- → **ARITHMETIC SEQUENCES**
- → **GEOMETRIC SEQUENCES**
- → **INFINITE GEOMETRIC SERIES**

A **sequence** is a function with a domain that consists of the natural numbers. A **series** is the sum of the terms of a sequence.

→ Examples

(A) an infinite sequence

$\frac{1}{2}, \frac{1}{3}, \frac{1}{4}, \frac{1}{5}, \ldots, \frac{1}{n+1}, \ldots,$ is an **infinite sequence** with

$t_1 = \frac{1}{2}, t_2 = \frac{1}{3}, t_3 = \frac{1}{4}, t_4 = \frac{1}{5}, \ldots, t_n = \frac{1}{n+1}.$

(B) a finite sequence

$2, 4, 6, \ldots, 20$ is a **finite sequence** with $t_1 = 2, t_2 = 4, t_3 = 6, \ldots, t_{10} = 20.$

(C) an infinite series

$\frac{1}{2} + \frac{1}{4} + \frac{1}{8} + \frac{1}{16} + \ldots + \frac{1}{2^n} + \ldots$ is an **infinite series**.

(D) If $t_n = \dfrac{2n}{n+1}$, **find the first five terms of the sequence.**

When 1, 2, 3, 4, and 5 are substituted for n, $t_1 = \frac{2}{2} = 1, t_2 = \frac{4}{3}, t_3 = \frac{6}{4} = \frac{3}{2}, t_4 = \frac{8}{5},$ and

$t_5 = \frac{10}{6} = \frac{5}{3}.$

The first five terms are $1, \frac{4}{3}, \frac{3}{2}, \frac{8}{5}, \frac{5}{3}.$

In Examples (A) through (D), formulas were given for finding each term of the sequence. Another way of specifying a sequence provides a formula that gives each term as a function of its predecessor terms. Examples (E) and (F) illustrate this method.

(E) If $a_1 = 3$ and $a_n = 2a_{n-1} + 5$, find a_4.

Put $3(a_1)$ into your graphing calculator, and press ENTER. Then multiply by 2 and add 5. Hit ENTER 3 more times to get $a_4 = 59$.

(F) If $a_1 = 1$, $a_2 = 1$, and $a_n = a_{n-1} + a_{n-2}$ for $n \geq 3$, find the first 7 terms of the sequence.

The recursive formula indicates that each term is the sum of the two terms before it. Therefore, the first seven terms of this sequence are $1, 1, 2, 3, 5, 8, 13$. This is called the Fibonacci sequence.

A **series** is the sum of the terms in a sequence. The expression $\sum\limits_{k=1}^{n} t_k$ represents the sum of n terms beginning with t_1. The lower limit of the sum shows the starting point, in this case 1, and the upper limit is the ending point, in this case n. Each t_k represents the kth term of the sequence ($k = 1, 2, \ldots, n$).

(G) Express the series $2 + 4 + 6 + \ldots + 20$ in sigma notation.

The series $2 + 4 + 6 + \ldots + 20 = \sum\limits_{i=1}^{10} 2i = 110$.

(H) Evaluate $\sum\limits_{k=0}^{5} k^2$.

$$\sum_{k=0}^{5} k^2 = 0^2 + 1^2 + 2^2 + 3^2 + 4^2 + 5^2 = 0 + 1 + 4 + 9 + 16 + 25 = 55$$

One of the most common sequences studied at this level is an **arithmetic sequence** (or **arithmetic progression**). Each term differs from the preceding term by a common difference. The first n terms of an arithmetic sequence can be denoted by the following:

$$t_1, t_1 + d, t_1 + 2d, t_1 + 3d, \ldots, t_1 + (n-1)d$$

d is the common difference and $t_n = t_1 + (n-1)d$. The sum of n terms of the series constructed from an arithmetic sequence is given by the formula:

$$S_n = \frac{n}{2}(t_1 + t_n) \quad \text{or} \quad S_n = \frac{n}{2}[2t_1 + (n-1)d]$$

(I) Find the 28th term of the arithmetic sequence 2, 5, 8, ...

$$t_n = t_1 + (n-1)d$$
$$t_1 = 2,\ d = 3,\ n = 28$$
$$t_{28} = 2 + 27 \cdot 3 = 83$$

(J) Express the sum of 28 terms of the series of this sequence using sigma notation.

$$\sum_{k=0}^{27}(3k+2) \ \text{ or } \ \sum_{j=1}^{28}(3j-1)$$

(K) Find the sum of the first 28 terms of the series.

$$S_n = \frac{n}{2}(t_1 + t_n)$$
$$S_{28} = \frac{28}{2}(2+83) = 14 \cdot 85 = 1,190$$

(L) If $t_8 = 4$ and $t_{12} = -2$, find the first three terms of the arithmetic sequence.

$$t_n = t_1 + (n-1)d$$
$$4 = t_1 + 7d$$
$$-2 = t_1 + 11d$$

To solve these two equations for d, subtract the second equation from the third.

$$-6 = 4d$$

$$d = -\frac{3}{2}$$

Substituting in the first equation gives $4 = t_1 + 7\left(-\dfrac{3}{2}\right)$. Thus:

$$t_1 = 4 + \frac{21}{2} = \frac{29}{2}$$
$$t_2 = \frac{29}{2} + \left(-\frac{3}{2}\right) = \frac{26}{2} = 13$$
$$t_3 = \frac{29}{2} + 2\left(-\frac{3}{2}\right) = \frac{23}{2}$$

The first three terms are $\dfrac{29}{2}, 13, \dfrac{23}{2}$.

(M) In an arithmetic series, if $S_n = 3n^2 + 2n$, find the first three terms.

When $n = 1$, $S_1 = t_1$. Therefore, $t_1 = 3(1)^2 + 2 \cdot 1 = 5$.

$$
\begin{aligned}
S_2 = t_1 + t_2 &= 3(2)^2 + 2 \cdot 2 = 16 \\
5 + t_2 &= 16 \\
t_2 &= 11
\end{aligned}
$$

Therefore, $d = 6$, which leads to a third term of 17. Thus, the first three terms are 5, 11, 17.

Another common type of sequence studied at this level is a **geometric sequence** (or geometric progression). In a geometric sequence the ratio of any two successive terms is a constant r called the constant ratio. The first n terms of a geometric sequence can be denoted by:

$$
t_1, \, t_1 r, \, t_1 r^2, \, t_1 r^3, \, \ldots, \, t_1 r^{n-1} = t_n
$$

The sum of the first n terms of a geometric series is given by the formula:

$$
S_n = \frac{t_1 \left(1 - r^n\right)}{1 - r}
$$

➡ **Examples**_____

(N) Find the seventh term of the geometric sequence $1, 2, 4, \ldots$.

$$
r = \frac{t_2}{t_1} = \frac{2}{1} = 2; \quad t_7 = t_1 r^{7-1}; \quad t_7 = 1 \cdot 2^6 = 64
$$

(O) Find the sum of the first seven terms of the geometric sequence $1, 2, 4, \ldots$.

$$
S_7 = \frac{1\left(1 - 2^7\right)}{1 - 2} = \frac{1 - 128}{-1} = 127
$$

(P) The first term of a geometric sequence is 64, and the common ratio is $\frac{1}{4}$.

For what value of n is $t_n = \frac{1}{4}$?

$$
\frac{1}{4} = 64 \left(\frac{1}{4}\right)^{n-1}
$$

$$
\left(\frac{1}{4}\right)^{2-n} = 64 = 4^3
$$

$$
4^{n-2} = 4^3
$$

$$
n = 5
$$

In a geometric sequence, if $|r| < 1$, the sum of the series approaches a limit as n approaches infinity. In the formula $S_n = \dfrac{t_1\left(1-r^n\right)}{1-r}$, if $|r| < 1$, the term $r^n \to 0$ as $n \to \infty$. Therefore, as long as $|r| < 1$, $\lim\limits_{n\to\infty} S_n = \dfrac{t_1}{1-r}$, or $S = \dfrac{t_1}{1-r}$.

➥ Examples

(Q) Evaluate $\lim\limits_{n\to\infty} \sum\limits_{k=1}^{n} \dfrac{1}{2^k}$

When the first few terms, $\dfrac{1}{2}+\dfrac{1}{4}+\dfrac{1}{8}+\ldots$, are listed, it can be seen that $t_1 = \dfrac{1}{2}$ and the common ratio $r = \dfrac{1}{2}$. Therefore:

$$S = \frac{\dfrac{1}{2}}{1-\dfrac{1}{2}} = 1$$

(R) Evaluate $\sum\limits_{j=0}^{\infty} (-3)^{-j}$

When the first few terms, $\dfrac{1}{1}-\dfrac{1}{3}+\dfrac{1}{9}-\ldots$, are listed, it can be seen that $t_1 = 1$ and the common ratio $r = -\dfrac{1}{3}$. Therefore:

$$S = \frac{1}{1-\left(-\dfrac{1}{3}\right)} = \frac{1}{\dfrac{4}{3}} = \frac{3}{4}$$

(S) Find the exact value of the repeating decimal 0.4545 . . . .

This can be represented by a geometric series, $0.45 + 0.0045 + 0.000045 + \ldots$, with $t_1 = 0.45$ and $r = 0.01$. Since $|r| < 1$:

$$S = \frac{0.45}{1-0.01} = \frac{0.45}{0.99} = \frac{45}{99} = \frac{5}{11}$$

1. If $a_1 = 3$ and $a_n = n + a_{n-1}$, the sum of the first five terms is

   (A) 17
   (B) 30
   (C) 42
   (D) 45
   (E) 68

2. If $a_1 = 5$ and $a_n = 1 + \sqrt{a_{n-1}}$, find a_3.

   (A) 2.623
   (B) 2.635
   (C) 2.673
   (D) 2.799
   (E) 3.323

3. If the repeating decimal $0.237\overline{37}\ldots$ is written as a fraction in lowest terms, the sum of the numerator and the denominator is

   (A) 16
   (B) 47
   (C) 245
   (D) 334
   (E) 1,237

4. The first three terms of a geometric sequence are $\sqrt[4]{3}, \sqrt[8]{3}, 1$. The fourth term is

   (A) $\sqrt[32]{3}$

   (B) $\sqrt[16]{3}$

   (C) $\dfrac{1}{\sqrt[16]{3}}$

   (D) $\dfrac{1}{\sqrt[8]{3}}$

   (E) $\dfrac{1}{\sqrt[4]{3}}$

5. In an infinite geometric series, $S = \dfrac{2}{3}$ and $t_1 = \dfrac{2}{7}$. What is r?

   (A) $\dfrac{2}{3}$

   (B) $-\dfrac{4}{7}$

   (C) $\dfrac{2}{7}$

   (D) $\dfrac{4}{7}$

   (E) $-\dfrac{2}{7}$

Answers Explained

1. **(D)** $a_2 = 5, a_3 = 8, a_4 = 12, a_5 = 17$. Therefore, $S_5 = 45$.

2. **(D)** Press 5 ENTER into your graphing calculator. Then enter $1 + \sqrt{\text{Ans}}$ and press ENTER twice more to get a_3.

3. **(C)** The decimal $0.23\overline{37} = 0.2 + (0.037 + 0.00037 + 0.0000037 + \ldots)$, which is $0.2 +$ an infinite geometric series with a common ratio of 0.01.

$$S_n = 0.2 + \frac{0.037}{0.99} = \frac{2}{10} + \frac{37}{990} = \frac{235}{990} = \frac{47}{198}$$

The sum of the numerator and the denominator is 245.

4. **(D)** The first three terms are $3^{1/4}, 3^{1/8}, 1$. Common ratio $= 3^{-1/8}$. Therefore, the fourth term is

$1 \cdot 3^{-1/8} = 3^{-1/8}$ or $\dfrac{1}{\sqrt[8]{3}}$.

5. **(D)** $\dfrac{2}{3} = \dfrac{\frac{2}{7}}{1-r}$. $2 - 2r = \dfrac{6}{7}$. $14 - 14r = 6$. Therefore, $r = \dfrac{4}{7}$.

Vectors

20

→ **DEFINITION**
→ **RESULTANT**
→ **NORM OF A VECTOR**
→ **DOT PRODUCT**

A vector in a plane is defined to be an ordered pair of real numbers. A vector in space is defined as an ordered triple of real numbers. On a coordinate system, a vector is usually represented by an arrow whose initial point is the origin and whose terminal point is at the ordered pair (or triple) that names the vector. Vector quantities always have a magnitude or **norm** (the length of the arrow) and direction (the angle the arrow makes with the x-axis). Vectors are often used to represent motion or force.

Properties of two-dimensional vectors can be extended to three-dimensional vectors. We will express the properties in terms of two-dimensional vectors for convenience. If vector $\vec{V}$ is designated by (v_1, v_2) and vector $\vec{U}$ is designated by (u_1, u_2), vector $\overrightarrow{U+V}$ is designated by $(u_1 + v_1, u_2 + v_2)$ and called the **resultant** of $\vec{U}$ and $\vec{V}$. Vector $-\vec{V}$ has the same magnitude as $\vec{V}$ but has a direction opposite that of $\vec{V}$.

On the plane, every vector $\vec{V}$ can be expressed in terms of any other two unit (magnitude 1) vectors parallel to the x- and y-axes. If vector $\vec{i} = (1, 0)$ and vector $\vec{j} = (0, 1)$, then vector $\vec{V} = ai + bj$, where a and b are real numbers, is the same as vector (a, b). A unit vector parallel to $\vec{V}$ can be determined by dividing $\vec{V}$ by its **norm**, denoted by $\|\vec{V}\|$ and equal to $\sqrt{a^2 + b^2}$.

It is possible to determine algebraically whether two vectors are perpendicular by defining the **dot product** or **inner product** of two vectors, $\vec{V}(v_1, v_2)$ and $\vec{U}(u_1, u_2)$.

$$\vec{V} \cdot \vec{U} = v_1 u_1 + v_2 u_2$$

Notice that the dot product of two vectors is a **real number**, not a vector. Two vectors, $\vec{V}$ and $\vec{U}$, are perpendicular if and only if $\vec{V} \cdot \vec{U} = 0$.

➡ Examples

(A) Let vector $\vec{V} = (2, 3)$ and vector $\vec{U} = (6, -4)$.

(i) What is the resultant of $\vec{U}$ and $\vec{V}$?

The resultant, $\overrightarrow{U+V}$, equals $(6 + 2, -4 + 3) = (8, -1)$.

(ii) What is the norm of $\vec{U}$?

The norm of $\vec{U}$, $\|\vec{U}\| = \sqrt{36+16} = \sqrt{52} = 2\sqrt{13}$.

(iii) Express $\vec{V}$ in terms of $\vec{i}$ and $\vec{j}$.

$\vec{V} = 2\vec{i} + 3\vec{j}$. To verify this, use the definitions of $\vec{i}$ and $\vec{j} \cdot \vec{V} = 2(1, 0) + 3(0, 1) = (2, 0) + (0, 3) = (2, 3) = \vec{V}$.

(iv) Are $\vec{U}$ and $\vec{V}$ perpendicular?

$\vec{U} \cdot \vec{V} = 6 \cdot 2 + (-4) \cdot 3 = 12 - 12 = 0$. Therefore, $\vec{U}$ and $\vec{V}$ are perpendicular because the dot product is equal to zero.

(B) If $\vec{U} = (-1, 4)$ and the resultant of $\vec{U}$ and $\vec{V}$ is $(4, 5)$, find $\vec{V}$.

Let $\vec{V} = (v_1, v_2)$. The resultant $\overrightarrow{U + V} = (-1, 4) + (v_1, v_2) = (4, 5)$. Therefore, $(-1 + v_1, 4 + v_2) = (4, 5)$, which implies that $-1 + v_1 = 4$ and $4 + v_2 = 5$. Thus, $v_1 = 5$ and $v_2 = 1$. $\vec{V} = (5, 1)$.

EXERCISES

1. Suppose $\vec{x} = (-3, -1)$, $\vec{y} = (-1, 4)$. Find the magnitude of $\vec{x} + \vec{y}$.

   (A) 2
   (B) 3
   (C) 4
   (D) 5
   (E) 6

2. If $\vec{V} = 2\vec{i} + 3\vec{j}$ and $\vec{U} = \vec{i} - 5\vec{j}$, the resultant vector of $2\vec{U} + 3\vec{V}$ equals

   (A) $3\vec{i} - 2\vec{j}$

   (B) $5\vec{i} + \vec{j}$

   (C) $7\vec{i} - 9\vec{j}$

   (D) $8\vec{i} - \vec{j}$

   (E) $2\vec{i} + 3\vec{j}$

3. A unit vector that is perpendicular to vector $\vec{V} = (3, -4)$ is

    (A) $(4, 3)$

    (B) $\left(\dfrac{3}{5}, \dfrac{4}{5}\right)$

    (C) $\left(-\dfrac{3}{5}, -\dfrac{4}{5}\right)$

    (D) $\left(-\dfrac{4}{5}, -\dfrac{3}{5}\right)$

    (E) $\left(-\dfrac{4}{5}, \dfrac{3}{5}\right)$

4. Given $\vec{V} = 2\vec{i} - 4\vec{j}$ and $\vec{V} = -6\vec{i} - \vec{j}$, the dot product of $\vec{U}$ and $\vec{V}$ is

    (A) -26
    (B) -16
    (C) -8
    (D) 0
    (E) 22

5. The norm of the vector $\vec{U} = (9, -4)$ is

    (A) $\sqrt{5}$
    (B) 5
    (C) $\sqrt{97}$
    (D) 10
    (E) 13

Answers Explained

1. **(D)** Add the components to get $\vec{x} + \vec{y} = (-4, 3)$. The magnitude is $\sqrt{(-4)^2 + 3^2} = 5$.

2. **(D)** $2\vec{U} = 2\vec{i} - 10\vec{j}$ and $3\vec{V} = 6\vec{i} + 9\vec{j}$, so $2\vec{U} + 3\vec{V} = 8\vec{i} - \vec{j}$.

3. **(D)** All of the answer choices except Choice A are unit vectors. Backsolve to find that the only one that has a zero dot product with $(3, -4)$ is $\left(-\dfrac{4}{5}, -\dfrac{3}{5}\right)$.

4. **(C)** The dot product is $(2)(-6) + (-4)(-1) = -8$.

5. **(C)** The norm of this vector is $\sqrt{9^2 + (-4)^2} = \sqrt{97}$.

Statistics

21

→ **MEAN, MEDIAN, MODE**
→ **RANGE AND STANDARD DEVIATION**
→ **z-SCORE**
→ **LINEAR, QUADRATIC, AND EXPONENTIAL MODELS**

Quantitative data are number sets such as heights, weights, test scores, tensile strength, and so forth. By contrast, **categorical** data consist of descriptive labels, such as hair color, city of residence, socioeconomic status, and the like. Since the Level 2 Mathematics Subject Test is unlikely to include questions about categorical data, the concepts described below pertain to quantitative data only.

Measures of center summarize a data set using a single "typical" value. Three measures of center might be encountered on the Level 2 Mathematics Subject Test: mean, median, and mode.

The **mean** is the sum of all the data values divided by the number of values. The formula for the mean $\bar{x}$ of a data set is $\bar{x} = \dfrac{\sum x_i}{n}$, where $\sum$ indicates the sum of the data values x_i and n is the number of data values.

To determine the **median**, the data must first be ordered. If the number of values is odd, the median is the single middle value. If the number of values is even, the median is the mean of the two middle values. There is no formula for the median of a data set.

The **mode** is the value that appears most often. There is no formula for the mode of a data set.

➥ Examples

(A) The heights of the starting basketball team for South High School are 69″, 72″, 75″, 78″, and 78″. Find the mean, median, and mode of this data set.

The mean is $\dfrac{69 + 72 + 75 + 78 + 78}{5} = 74.4''$. The median is $75''$. The mode is $78''$.

(B) The mean of 24 test scores is 77.5. When the 25th class member takes the test, the mean goes down by 1.1 points. What was that 25th score?

The total of the 24 test scores is $24 \times 77.5 = 1,860$, and the total of the 25 test scores is $25 \times 76.4 = 1,910$. Therefore, the 25th score is $1,910 - 1,860 = 50$.

(C) What is the median of the frequency distribution shown in the following table?

Data Value	Frequency
24	3
25	7
26	5
27	1

There are 16 data values altogether, so the median is the mean of the 8th and 9th largest values. Both of these values are 25, so the median is also 25.

A **weighted average** is another way of measuring the middle of a data set. The data consist of numbers (often averages themselves) for various categories that are to be combined to form an overall average. Higher weights are given to categories that are considered more important. To get the weighted overall average, the number in each category is then multiplied by the category's weight, and these products are summed across the categories.

➡ **Example** _____

(D) **In some schools, AP courses, honors courses, and college prep courses carry different weights in the computation of grade point averages. A given grade may carry a weight of 2 in an AP course, a weight of 1.5 in an honors course, and a weight of 1 in a college prep course. Suppose letter grades of A, B, C, D, and F are assigned values of 4, 3, 2, 1, and 0.**

A student completes 6 courses with grades of a B and a C in 2 AP courses, an A and a C in 2 honors courses, and 2 As in college prep courses. What would that student's grade point average be?

The student's grade point average is calculated as the weighted average:

$$[2(3 + 2) + 1.5(4 + 2) + 1(4 + 4)] \div 6 = 4.5$$

The Level 2 Mathematics Subject Test might ask questions about **measures of spread**. These questions ask about how spread out a set of data values is.

The **range** is a measure of spread. It is the difference between the largest and smallest data values.

➡ **Example** _____

(E) **Find the range of the data values 85, 96, 72, 89, 66, and 78.**

The largest value is 96 and the smallest is 66. The range is $96 - 66 = 30$.

Loosely speaking, the **standard deviation** is the "average" difference between individual data values and their mean. The formula for the standard deviation s of a data set is $s = \sqrt{\frac{1}{n-1}\sum(x_i - \bar{x})^2}$. The larger the standard deviation, the more spread out a data set is.

Standard deviation is a unit-free measure of the "distance" between a specific data value and the mean. Thus the standard deviation can be used to compare single data values from different data sets. A z-score, where $z = \dfrac{x - \bar{x}}{s}$, is the number of standard deviations s that a data value x is from the mean $\bar{x}$. The greater the value of $|z|$, the less common the data value x is. In other words, fewer data values have a high z-score.

➡ **Examples** _____

(F) **Which data set has the smaller standard deviation: {5, 7, 9} or {4, 7, 10}?**

Both data sets have a mean of 7. However, the first set is less spread out than the second, so the first has the smaller standard deviation. According to the formula, the standard deviation of the first data set is 2 while that of the second data set is 3.

(G) **A chart that shows sports statistics for a particular school is shown below. Which is statistically a better score: 50.30 seconds in the backstroke or 74 inches in the high jump?**

Stroke	Mean	Standard Deviation
Backstroke	50.72 sec.	0.24 sec.
High Jump	72.9 in.	0.54 in.

A time of 50.30 seconds in the backstroke is $z = \dfrac{50.30 - 50.72}{0.24} = -1.75$ standard deviations better

(less) than the backstroke mean. A height of 74 inches in the high jump is $z = \dfrac{74 - 72.9}{0.54} = 2.04$

standard deviations better (more) than the high jump mean. Therefore, the high jump performance is better.

Measures of center and spread apply to a single variable. **Regression** is a technique for analyzing the relationship between two variables. This technique summarizes relationships such as mathematical equations in which the two variables are denoted by x (the independent variable) and y (the dependent variable). The Level 2 Mathematics Subject Test may ask about any one of three models to capture the relationship between x and y:

- Linear model: $y = a_0 + a_1 x$
- Quadratic model: $y = a_0 + a_1 x + a_2 x^2$
- Exponential model: $y = a_0 e^{a_1 x}$

The first step when deciding which model should be used in a particular situation is to examine a scatterplot of the data. Enter the data into two lists (STAT/Edit). Clear the Y= screen, key 2nd Y=, and select a plot. This takes you to the Statplot screen. Turn the plot on, select the scatterplot image, enter the list names, and key ZOOM/Statplot.

If the points in the scatterplot fall roughly in a straight line, enter STAT/CALC/LinReg (either version). The calculator will return a linear equation. Linear models are appropriate when the dependent variable changes at a constant rate relative to the independent variable.

If the points seem to change direction once, a quadratic model is probably appropriate. Enter STAT/CALC/QuadReg, and the calculator will return a quadratic equation in standard form. Quadratic models apply when the dependent variable varies with the square of the independent variable. Examples are (a) the area of a figure varies with the square of the length of its boundary, and (b) the intensity of light varies with the square of its distance from the light source.

If the scatterplot is curved but does not change direction, an exponential model is probably the best choice. The command on the graphing calculator is STAT/CALC/ExpReg. Exponential models apply when the rate of change in the dependent variable is proportional to the independent variable. Examples are population growth, the spread of disease, and nuclear decay.

The figures below show scatterplots that have these shapes. Regression techniques use paired values (x, y) to estimate **parameter** values a_0, a_1, a_2, depending on the model selected. Once this is done, the equation for that model can be used to predict y for a given value of x.

Linear Scatterplot

Quadratic Scatterplot

Exponential Scatterplot

The Level 2 Mathematics Subject Test does not require students to know the mathematics of regression techniques. Students should know how to use their calculators to get parameter estimates for a particular model and how to use the equation as a prediction tool.

➥ **Example**

(H) **The decennial population of Center City for the past five decades is shown in the table below. Use exponential regression to estimate the 1965 population.**

Population of Center City

Year	Population
1950	48,000
1960	72,000
1970	95,000
1980	123,000
1990	165,000

Transform the years to "number of years after 1950" and enter these values into L4. Then enter the populations in thousands. Set up the scatterplot by pressing 2ndY= and selecting a plot (Plot 1). Turn the plot on, select the scatterplot logo, and enter the list names. Then press STAT/CALC/ExpReg L1,L2,blank,Y_1,Calculate. This will store the regression equation in Y_1. The resulting command is shown in the left screen below. Press ENTER to display the values for the equation. These are shown in the right screen below.

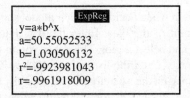

Press ZOOM/9 to view the scatterplot and exponential curve. Press 2nd/CALC/value and enter 15, representing 1965. The cursor moves to the point on the regression curve where $x = 15$ and displays both x and y at the bottom of the screen, as shown below.

The 1965 population was about 79,300.

1. Last week, police ticketed 13 men traveling 18 miles per hour over the speed limit and 8 women traveling 14 miles per hour over the speed limit. What was the mean over the speed limit of all 21 drivers?

   (A) 16 miles per hour
   (B) 16.5 miles per hour
   (C) 17 miles per hour
   (D) none of the above
   (E) cannot be determined

2. Wanda wants to choose her "best-fit" college in a systematic way. Academic reputation (AR), sports reputation (SR), social life (SL), and cost (C) are the four factors that Wanda considers for all her colleges. She plans to score the colleges on these factors using a scale of 1 to 5, with 5 being the highest, and she attaches the following weights: AR (40%), SR (10%), SL (15%), and C (35%). Her scores for these categories at one school are AR, 4; SR, 3; SL, 5; C, 2. What is the weighted overall average score for this school?

   (A) 3.35
   (B) 3.5
   (C) 3.85
   (D) 4.25
   (E) 4.5

3. If the range of a set of integers is 2 and the mean is 50, which of the following statements must be true?

   I. The mode is 50.
   II. The median is 50.
   III. There are exactly three data values.

   (A) only I
   (B) only II
   (C) only III
   (D) I and II
   (E) I, II, and III

4. What is the median of the frequency distribution shown below?

Data Value	Frequency
0	1
1	3
2	7
3	15
4	10
5	7
6	3
7	3

   (A) 2
   (B) 3
   (C) 4
   (D) 5
   (E) cannot be determined

5. Which of the following statements must be true?

   I. The range of a data set must be smaller than its standard deviation.
   II. The standard deviation of a data set must be smaller than its mean.
   III. The median of a data set must be smaller than its mode.

(A) I only
(B) I and II
(C) II only
(D) I, II, and III
(E) none are true

6. The mean and standard deviation for SAT math scores are shown in the table below for five high schools in a large city. A particular score for each school is also shown (in the right column).

School	Mean	Standard Deviation	Single Score
A	532	24	600
B	485	30	560
C	515	22	561
D	396	26	474
E	479	35	552

Which single score has the highest z-score?

(A) 474 in school D
(B) 552 in school E
(C) 560 in school B
(D) 561 in school C
(E) 600 in school A

7. Jack recorded the amount of time he studied the night before each of 4 history quizzes and the score he got on each quiz. The data are in the table below.

Score	Time (min.)
86	45
70	15
90	40
78	35

Use linear regression to estimate the score Jack would get if he studied for 20 minutes.

(A) 71
(B) 72
(C) 73
(D) 74
(E) 75

8. The scatterplot below shows gas mileage (miles per gallon) at various speeds (miles per hour) when a car was driven 100 miles at various speeds on a test track.

Speed (mph)

Which regression model is probably the best predictor of gas mileage as a function of speed?

(A) constant
(B) linear
(C) quadratic
(D) cubic
(E) exponential

Answers Explained

✱ 1. **(B)** There are 13 eighteens and 8 fourteens, so the total over the speed limit is 346. Divide this by the 21 people to get 16.5.

✱ 2. **(A)** The score in each category is multiplied by that category's weight, and these products are added to get $0.4(4) + 0.1(3) + 0.15(5) + 0.35(2) = 3.35$ as the weighted overall average score for that school.

3. **(B)** Since the data values are integers, the range is 2, and the mean is 50, the possible data values are 49, 50, and 51.

    I. The set could consist of equal numbers of 49s and 51s and have a mean of 50 without 50 even being a data value. So I need not be true.

    II. Since the mean is 50, there must be equal numbers of 49s and 51s, so 50 is also the median. II must be true.

    III. The explanations in I and II imply that III need not be true.

4. **(B)** There are 49 data values altogether, so the median is the 25th largest. Adding the frequencies up to 25 puts the 25th number at 3.

5. **(E)** None are true. The range of any data set must be larger than its standard deviation because the range measures total spread while the standard deviation measures average spread. So I is false. Either the mean or the standard deviation of a data set can be larger. For example, the mean of the data $\{1, 5, 10\}$ is 5.3, while its standard deviation is 4.51. The mean of the data set $\{1, 5, 20\}$ is 8.7, while its standard deviation is 10.0. So II is false. Either the median or mode of a data set can be larger. For example, the median of the data set $\{1, 2, 3, 4, 4\}$ is 3, while its mode is 4. The median of the data set $\{1, 1, 2, 3, 4\}$ is 2 while its mode is 1. So III is false.

✱ 6. **(A)** The z-scores for the five schools are 2.8 for A, 2.5 for B, 2.1 for C, 3 for D, and 2.1 for E.

* 7. **(B)** Enter the data in two lists (study times in L1 and test scores in L2). Enter STAT/CALC/8, and enter VARS/YVARS/Function/Y_1, followed by ENTER. This produces estimates of the slope (b) and y-intercept (a) of the regression line $a + bx$. Enter this expression into Y_1. Enter $Y_1(2)$ to get the score of 72.

* 8. **(C)** The scatterplot has the shape of a parabola with a maximum. Therefore, the quadratic model would be the best predictor.

Probability

22

→ **SAMPLE SPACE**
→ **MUTUALLY EXCLUSIVE EVENTS**
→ **CONDITIONAL PROBABILITY AND INDEPENDENT EVENTS**

When an "experiment" has an uncertain outcome, probability provides a measure of how likely an outcome or collection of outcomes is. Tossing a coin, rolling dice, drawing cards from a standard deck, and picking marbles out of bowls are simple examples of experiments that illustrate important aspects of probability.

The results of an experiment are called its outcomes. A coin toss has two outcomes: heads (H) and tails (T). Tossing two coins has four (2 × 2) outcomes: (HH), (HT), (TH), and (TT). Rolling a single die has six possible outcomes, while rolling a pair of dice has thirty-six (6 × 6) possible outcomes. Drawing a single card from a standard deck has 52 possible outcomes. Drawing five cards from a standard deck has $\binom{52}{5} = 2,598,960$ possible outcomes (five-card poker hands).

The set of all possible outcomes of some experiment is called its **sample space**. An **event** is a subset of the sample space (a collection of possible outcomes). The **probability of an event** is the number of outcomes in the event (**favorable outcomes**) divided by the number of outcomes in the sample space.

Let E be an event, $\widetilde{E}$ be the set of outcomes that are not in E (the **complement** of E), and $P(E)$ represent the probability of E.

- $0 \leq P(E) \leq 1$
- $P(\widetilde{E}) = 1 - P(E)$

If $P(E) = 0$, E is called the **impossible event**. If $P(E) = 1$, E is called the **certain event**.

➡ Examples

(A) If a card is drawn from a standard deck of 52 cards, what is the probability of the event that the card is a picture card? What is the probability of the event that the card is a spade?

There are 12 picture cards in a standard deck, so if one card is drawn, the probability that it is a picture card is $\frac{12}{52}$. The probability that it is a spade is $\frac{13}{52}$ because a deck has 13 spades.

(B) If a coin is flipped twice, what is the probability of the event 1 head and 1 tail?

Since two of the four possible outcomes have one head and one tail, the probability of this event is $\frac{2}{4} = \frac{1}{2}$.

(C) What is the probability of getting a sum of 7 when two dice are thrown?

The outcomes of the rolling of a pair of dice can be more readily understood by assigning colors to the dice (e.g., one red, one green). An outcome of throwing a pair of dice can be represented by an ordered pair, where the first element is the side of the red die facing up, and the second is the side of the green die facing up. Thus, (32) signifies a 3 on the red die and a 2 on the green, while (23) signifies 2 on the red and 3 on the green. Since there are six faces on each die, there are $6 \times 6 = 36$ possible outcomes in the sample space. The outcomes that have a sum of 7 are (61), (16), (52), (25), (43), and (34). Therefore, the probability of getting a sum of 7 is $\dfrac{6}{36} = \dfrac{1}{6}$.

Probability questions can also be asked about two events A and B. Questions on the Level 2 Mathematics Subject Test cover the following three ways of combining two events.

- $P(A \cup B)$, the probability of either A *or* B occurring, where $A \cup B$ is called the **disjunction** of A and B.
- $P(A \mid B)$, the probability of A occurring, *given* that B has occurred. This is called **conditional probability**.
- $P(A \cap B)$, the probability of both A *and* B occurring, where $A \cap B$ is called the **conjunction** of A and B.

If two events have no outcomes in common, they are called **mutually exclusive**, and $P(A \cup B) = P(A) + P(B)$. If two events have outcomes in common, these outcomes would be double counted, resulting in the more general formula $P(A \cup B) = P(A) + P(B) - P(A \cap B)$. Note that A and B are mutually exclusive if and only if $P(A \cap B) = 0$.

➡️ **Examples**_____

(D) In a group of 40 students, 23 take the AP Psychology class, 18 take the AP Calculus class, and 8 take both classes. What is the probability that a student takes AP Psychology or AP Calculus?

Let A be the event that a student takes AP Psychology and B be the event that a student takes AP Calculus. Then $A \cap B$ represents the event that a student takes both courses. In the formula above, $P(A) = \dfrac{23}{40}$, $P(B) = \dfrac{18}{40}$, and $P(A \cap B) = \dfrac{8}{40}$. Therefore:

$$P(A \cup B) = \frac{23}{40} + \frac{18}{40} - \frac{8}{40} = \frac{33}{40}$$

Taking both AP Psychology and AP Calculus are *not* mutually exclusive because the probability of taking both courses is not zero.

(E) In a throw of two dice, what is the probability of getting a 7 or 11?

Let A be the event of throwing a 7 and B be the event of throwing (65) or (56) for a total of 11. Since these events are mutually exclusive $P(A \cup B) = P(A) + P(B) = \dfrac{6}{36} + \dfrac{2}{36} = \dfrac{8}{36}$.

In **conditional probability**, the sample space is reduced to the set B rather than the set of all outcomes, and the favorable outcomes are those in both A and B ($A \cap B$). $P(A \mid B)$ is the probability that A occurs given that B occurred. This results in the formula $P(A \mid B) = \dfrac{P(A \cap B)}{P(B)}$.

Multiplying both sides of this equation by $P(B)$ produces the equivalent formula $P(A \cap B) = P(B)P(A \mid B)$. If the probability of event A occurring does not depend on whether event B occurs, then A and B are said to be **independent** and $P(A \mid B) = P(A)$. Thus, if A and B are independent events, $P(A \cap B) = P(B)P(A)$.

(F) **If you draw two marbles in sequence from a bowl with 10 black and 4 yellow marbles and replace the first marble before picking the second, what is the probability that both marbles are yellow?**

Since there are 14 marbles altogether, and 4 are yellow, the probabilities for the first marble are $\frac{4}{14}$ for Y and $\frac{10}{14}$ for B. The return of the first marble makes the first and second picks independent, so the second pick has the same probabilities. Use the formula

$P(A \cap B) = P(A)P(B)$, where A and B are both Y (yellow), to find:

$$P(Y \cap Y) = \frac{4}{14} \cdot \frac{4}{14} = \frac{16}{196}$$

(G) **The scenario is the same as in Example (F), but this time, the first marble picked is not replaced. What is the probability that both marbles are yellow?**

If the first marble is not returned, the first and second picks are not independent, so we must use the formula $P(A \cap B) = P(A \mid B)P(B)$. In this case, $P(Y) = \frac{4}{14}$ as before, but $P(Y \mid Y) = \frac{3}{13}$ because there are only 13 marbles left, of which only 3 are yellow. Therefore:

$$P(Y \cap Y) = P(Y)P(Y \mid Y) = \left(\frac{4}{14}\right)\left(\frac{3}{13}\right) = \frac{12}{182} = \frac{6}{91}$$

(H) **If the probability that John will buy a certain product is $\frac{3}{5}$, that Bill will buy that product is $\frac{2}{3}$, and that Sue will buy that product is $\frac{1}{4}$, and if their decisions to buy are independent, what is the probability that at least one of them will buy the product?**

This is a good example of using complements. Since the question asks for the probability of "at least one," find the probability of "none" and subtract that from 1. The probabilities that John, Bill, and Sue don't buy the product are $\frac{2}{5}, \frac{1}{3}$, and $\frac{3}{4}$, respectively. Since their decisions to buy (or equivalently, not to buy) are independent, the probability that none of them buys the product is $\left(\frac{2}{5}\right)\left(\frac{1}{3}\right)\left(\frac{3}{4}\right) = \frac{1}{10}$. Therefore, the probability that at least one of them buys the product is $1 - \frac{1}{10} = \frac{9}{10}$.

(I) **Two dice are thrown. Event A is "the sum of the two dice is 7" and Event B is "at least one die is a 6." Are A and B independent?**

From Example (C), $P(A) = \frac{6}{36}$. Event B includes (1 and 6), (2 and 6), . . . , (6 and 6), (6 and 1), (6 and 2), . . . , (6 and 5), so $P(B) = \frac{11}{36}$. Only the two throws (6 and 1) and (1 and 6) are in the event $A \mid B$, so $P(A \mid B) = \frac{2}{36}$. $P(B)P(A) = \left(\frac{11}{36}\right)\left(\frac{6}{36}\right) = \frac{11}{216} \neq \frac{2}{36}$.

Therefore, A and B are *not* independent.

1. With the throw of two dice, what is the probability that the sum will be a prime number?

   (A) $\dfrac{4}{11}$

   (B) $\dfrac{7}{18}$

   (C) $\dfrac{5}{12}$

   (D) $\dfrac{5}{11}$

   (E) $\dfrac{1}{2}$

2. If a coin is flipped and one die is thrown, what is the probability of getting either a head or a 4?

   (A) $\dfrac{1}{12}$

   (B) $\dfrac{1}{3}$

   (C) $\dfrac{5}{12}$

   (D) $\dfrac{7}{12}$

   (E) $\dfrac{2}{3}$

3. Three cards are drawn from an ordinary deck of 52 cards. Each card is replaced in the deck before the next card is drawn. What is the probability that at least one of the cards will be a spade?

   (A) $\dfrac{3}{52}$

   (B) $\dfrac{9}{64}$

   (C) $\dfrac{3}{8}$

   (D) $\dfrac{37}{64}$

   (E) $\dfrac{3}{4}$

4. A coin is tossed three times. Let A = {three heads occur} and B = {at least one head occurs}. What is $P(A \cup B)$?

(A) $\dfrac{1}{8}$

(B) $\dfrac{1}{4}$

(C) $\dfrac{1}{2}$

(D) $\dfrac{3}{4}$

(E) $\dfrac{7}{8}$

5. A class has 12 boys and 4 girls. If three students are selected at random from the class, what is the probability that all will be boys?

(A) $\dfrac{1}{55}$

(B) $\dfrac{1}{4}$

(C) $\dfrac{1}{3}$

(D) $\dfrac{11}{28}$

(E) $\dfrac{11}{15}$

6. A red box contains eight items, of which three are defective, and a blue box contains five items, of which two are defective. An item is drawn at random from each box. What is the probability that both items will be non-defective?

(A) $\dfrac{3}{20}$

(B) $\dfrac{3}{8}$

(C) $\dfrac{5}{13}$

(D) $\dfrac{8}{13}$

(E) $\dfrac{17}{20}$

7. A hotel has five single rooms available, for which six men and three women apply. What is the probability that the rooms will be rented to three men and two women?

(A) $\dfrac{23}{112}$

(B) $\dfrac{97}{251}$

(C) $\dfrac{10}{21}$

(D) $\dfrac{5}{9}$

(E) $\dfrac{5}{8}$

8. Of all the articles in a box, 80% are satisfactory, while 20% are not. The probability of obtaining exactly five satisfactory articles out of eight randomly selected articles is

(A) 0.003
(B) 0.013
(C) 0.132
(D) 0.147
(E) 0.800

Answers Explained

1. **(C)** There is 1 way to get a 2, 2 ways to get a 3, 4 ways to get a 5, 6 ways to get a 7, and 2 ways to get an 11. Out of 36 elements in the sample space, 15 successes are possible.

$$P(\text{prime}) = \frac{15}{36} = \frac{5}{12}$$

2. **(D)** The probability of getting neither a head nor a 4 is $\dfrac{1}{2} \cdot \dfrac{5}{6} = \dfrac{5}{12}$. Therefore, the probability of getting either is $1 - \dfrac{5}{12} = \dfrac{7}{12}$.

* 3. **(D)** Since the drawn cards are replaced, the draws are independent. The probability that none of the cards was a spade is $\dfrac{39}{52} \cdot \dfrac{39}{52} \cdot \dfrac{39}{52} = \dfrac{3}{4} \cdot \dfrac{3}{4} \cdot \dfrac{3}{4} = \dfrac{27}{64}$.

The probability that one was a spade is $1 - \dfrac{27}{64} = \dfrac{37}{64}$.

4. **(E)** The only situation when neither of these sets is satisfied occurs when three tails appear. $P(A \cup B) = \dfrac{7}{8}$.

* 5. **(D)** There are 16 students altogether. The probability that the first person chosen is a boy is $\frac{12}{16}$. Now there are only 15 students left, of which 11 are boys, so the probability that the second student chosen is also a boy is $\frac{11}{15}$. By the same reasoning, the probability that the third is a boy is $\frac{10}{14}$. Therefore, the probability that the first, second, and third students chosen are all boys is $\frac{12}{16} \cdot \frac{11}{15} \cdot \frac{10}{14} = \frac{11}{28}$.

6. **(B)** The probability of both items being non-defective is $\frac{5}{8} \cdot \frac{3}{5} = \frac{3}{8}$.

* 7. **(C)** $\begin{pmatrix} 6 \\ 3 \end{pmatrix}$ is the number of ways 3 men can be selected. $\begin{pmatrix} 3 \\ 2 \end{pmatrix}$ is the number of ways 2 women can be selected. $\begin{pmatrix} 9 \\ 5 \end{pmatrix}$ is the total number of ways people can be selected to fill 5 rooms.

$$P(3 \text{ men}, 2 \text{ women}) = \frac{\begin{pmatrix} 6 \\ 3 \end{pmatrix}\begin{pmatrix} 3 \\ 2 \end{pmatrix}}{\begin{pmatrix} 9 \\ 5 \end{pmatrix}} = \frac{10}{21}$$

* 8. **(D)** Since the problem doesn't say how many articles are in the box, we must assume that it is an unlimited number. The probability of picking 5 satisfactory items (and therefore 3 unsatisfactory ones) is $(0.8)^5(0.2)^3$, and there are $\begin{pmatrix} 8 \\ 5 \end{pmatrix}$ ways of doing this. Therefore, the desired probability is $\begin{pmatrix} 8 \\ 5 \end{pmatrix}(0.8)^5(0.2)^3 \approx 0.147$.

PART 3
Practice Tests

ONLINE

Don't forget that in addition to the diagnostic test and the following six practice tests, Barron's *SAT Subject Test Math Level 2, 13th Edition* also includes two full-length online practice tests, which you can take at any time on your computer, tablet, or smartphone.

Visit *barronsbooks.com/tp/sat/math2/* to access the two additional online tests or scan the QR code below:

*Be sure to have your copy of *SAT Subject Test Math Level 2, 13th Edition* on hand to complete the registration process.

ANSWER SHEET
Practice Test 1

1. Ⓐ Ⓑ Ⓒ Ⓓ Ⓔ
2. Ⓐ Ⓑ Ⓒ Ⓓ Ⓔ
3. Ⓐ Ⓑ Ⓒ Ⓓ Ⓔ
4. Ⓐ Ⓑ Ⓒ Ⓓ Ⓔ
5. Ⓐ Ⓑ Ⓒ Ⓓ Ⓔ
6. Ⓐ Ⓑ Ⓒ Ⓓ Ⓔ
7. Ⓐ Ⓑ Ⓒ Ⓓ Ⓔ
8. Ⓐ Ⓑ Ⓒ Ⓓ Ⓔ
9. Ⓐ Ⓑ Ⓒ Ⓓ Ⓔ
10. Ⓐ Ⓑ Ⓒ Ⓓ Ⓔ
11. Ⓐ Ⓑ Ⓒ Ⓓ Ⓔ
12. Ⓐ Ⓑ Ⓒ Ⓓ Ⓔ
13. Ⓐ Ⓑ Ⓒ Ⓓ Ⓔ

14. Ⓐ Ⓑ Ⓒ Ⓓ Ⓔ
15. Ⓐ Ⓑ Ⓒ Ⓓ Ⓔ
16. Ⓐ Ⓑ Ⓒ Ⓓ Ⓔ
17. Ⓐ Ⓑ Ⓒ Ⓓ Ⓔ
18. Ⓐ Ⓑ Ⓒ Ⓓ Ⓔ
19. Ⓐ Ⓑ Ⓒ Ⓓ Ⓔ
20. Ⓐ Ⓑ Ⓒ Ⓓ Ⓔ
21. Ⓐ Ⓑ Ⓒ Ⓓ Ⓔ
22. Ⓐ Ⓑ Ⓒ Ⓓ Ⓔ
23. Ⓐ Ⓑ Ⓒ Ⓓ Ⓔ
24. Ⓐ Ⓑ Ⓒ Ⓓ Ⓔ
25. Ⓐ Ⓑ Ⓒ Ⓓ Ⓔ
26. Ⓐ Ⓑ Ⓒ Ⓓ Ⓔ

27. Ⓐ Ⓑ Ⓒ Ⓓ Ⓔ
28. Ⓐ Ⓑ Ⓒ Ⓓ Ⓔ
29. Ⓐ Ⓑ Ⓒ Ⓓ Ⓔ
30. Ⓐ Ⓑ Ⓒ Ⓓ Ⓔ
31. Ⓐ Ⓑ Ⓒ Ⓓ Ⓔ
32. Ⓐ Ⓑ Ⓒ Ⓓ Ⓔ
33. Ⓐ Ⓑ Ⓒ Ⓓ Ⓔ
34. Ⓐ Ⓑ Ⓒ Ⓓ Ⓔ
35. Ⓐ Ⓑ Ⓒ Ⓓ Ⓔ
36. Ⓐ Ⓑ Ⓒ Ⓓ Ⓔ
37. Ⓐ Ⓑ Ⓒ Ⓓ Ⓔ
38. Ⓐ Ⓑ Ⓒ Ⓓ Ⓔ
39. Ⓐ Ⓑ Ⓒ Ⓓ Ⓔ

40. Ⓐ Ⓑ Ⓒ Ⓓ Ⓔ
41. Ⓐ Ⓑ Ⓒ Ⓓ Ⓔ
42. Ⓐ Ⓑ Ⓒ Ⓓ Ⓔ
43. Ⓐ Ⓑ Ⓒ Ⓓ Ⓔ
44. Ⓐ Ⓑ Ⓒ Ⓓ Ⓔ
45. Ⓐ Ⓑ Ⓒ Ⓓ Ⓔ
46. Ⓐ Ⓑ Ⓒ Ⓓ Ⓔ
47. Ⓐ Ⓑ Ⓒ Ⓓ Ⓔ
48. Ⓐ Ⓑ Ⓒ Ⓓ Ⓔ
49. Ⓐ Ⓑ Ⓒ Ⓓ Ⓔ
50. Ⓐ Ⓑ Ⓒ Ⓓ Ⓔ

Practice Test 1

T ear out the preceding answer sheet. Decide which is the best choice by rounding your answer when appropriate. Blacken the corresponding space on the answer sheet. When finished, check your answers with those at the end of the test. For questions that you got wrong, note the chapters containing the material that you must review. Also, if you do not fully understand how you arrived at some of the correct answers, you should review the appropriate chapters. Finally, fill out the Self-Evaluation Chart on page 226 in order to pinpoint the topics that gave you the most difficulty.

50 questions: 1 hour

Directions: Decide which answer choice is best. If the exact numerical value is not one of the answer choices, select the closest approximation. Fill in the circle on the answer sheet that corresponds to your choice.

Notes:
(1) You will need to use a scientific or graphing calculator to answer some of the questions.
(2) You will have to decide whether to put your calculator in degree or radian mode for some problems.
(3) All figures that accompany problems are plane figures unless otherwise stated. Figures are drawn as accurately as possible to provide useful information for solving the problem, except when it is stated in a particular problem that the figure is not drawn to scale.
(4) Unless otherwise indicated, the domain of a function is the set of all real numbers for which the functional value is also a real number.

Reference Information. The following formulas are provided for your information.

Volume of a right circular cone with radius r and height h: $V = \dfrac{1}{3}\pi r^2 h$

Volume of a sphere of radius r: $V = \dfrac{4}{3}\pi r^3$

Surface area of a sphere of radius r: $S = 4\pi r^2$

Volume of a pyramid of base area B and height h: $V = \dfrac{1}{3}Bh$

1. The slope of a line perpendicular to the line whose equation is $\frac{x}{3} - \frac{y}{4} = 1$ is

(A) -3

(B) $-\frac{4}{3}$

(C) $-\frac{3}{4}$

(D) $\frac{1}{4}$

(E) $\frac{4}{3}$

2. What is the range of the data set 8, 12, 12, 15, 18?

(A) 10
(B) 12
(C) 13
(D) 15
(E) 18

3. If $f(x) = \frac{x-7}{x^2-49}$, for what value(s) of x does the graph of $y = f(x)$ have a vertical asymptote?

(A) -7
(B) 0
(C) $-7, 0, 7$
(D) $-7, 7$
(E) 7

4. If $f(x) = \sqrt{2x+3}$ and $g(x) = x^2 + 1$, then $f(g(2)) =$

(A) 2.24
(B) 3.00
(C) 3.61
(D) 6.00
(E) 6.16

5. What is the value of x if $4^{x+1} = 8^x$?

(A) -1
(B) 0
(C) 1
(D) 2
(E) 3

6. The circumference of circle $x^2 + y^2 - 10y - 36 = 0$ is

   (A) 38
   (B) 49
   (C) 54
   (D) 125
   (E) 192

7. Twenty-five percent of a group of unrelated students are only children. The students are asked one at a time whether they are only children. What is the probability that the 5th student asked is the first only child?

   (A) 0.00098
   (B) 0.08
   (C) 0.24
   (D) 0.25
   (E) 0.50

8. If $f(x) = 2$ for all real numbers x, then $f(x + 2) =$

   (A) 0
   (B) 2
   (C) 4
   (D) x
   (E) The value cannot be determined.

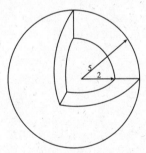

Figure not drawn to scale

9. The volume of the region between two concentric spheres of radii 2 and 5 is

   (A) 28
   (B) 66
   (C) 113
   (D) 368
   (E) 490

10. Find the value of $\log_b 12$ if $\log_b 2 = 0.4307$ and $\log_b 3 = 0.6826$.

   (A) 0.1266
   (B) 0.5879
   (C) 1.1133
   (D) 1.5440
   (E) 2.2266

11. In right triangle ABC, $AB = 10$, $BC = 8$, $AC = 6$. The sine of $\angle A$ is

(A) $\dfrac{3}{5}$

(B) $\dfrac{3}{4}$

(C) $\dfrac{4}{5}$

(D) $\dfrac{5}{4}$

(E) $\dfrac{4}{3}$

12. If $16^x = 4$ and $5^{x+y} = 625$, then $y =$

(A) 1

(B) 2

(C) $\dfrac{7}{2}$

(D) 5

(E) $\dfrac{25}{2}$

13. If the parameter is eliminated from the equations $x = t^2 + 1$ and $y = 2t$, then the relation between x and y is

(A) $y = x - 1$
(B) $y = 1 - x$
(C) $y^2 = x - 1$
(D) $y^2 = (x - 1)^2$
(E) $y^2 = 4x - 4$

14. Let $f(x)$ be a polynomial function: $f(x) = x^5 + \ldots$. If $f(1) = 0$ and $f(2) = 0$, then $f(x)$ is divisible by

(A) $x - 3$
(B) $x^2 - 2$
(C) $x^2 + 2$
(D) $x^2 - 3x + 2$
(E) $x^2 + 3x + 2$

15. The domain of $\dfrac{\sqrt{25 - x^2}}{\sqrt{x^2 - 9}}$ is

(A) $(-5,5)$
(B) $(-3,5)$
(C) $[-5,-3) \cup (3,5]$
(D) $(-5,-3) \cup (3,5)$
(E) $(-5,3) \cup (3,5)$

USE THIS SPACE FOR SCRATCH WORK

16. If $z > 0$, $a = z\cos\theta$, and $b = z\sin\theta$, then $\sqrt{a^2 + b^2} =$

   (A) 1
   (B) z
   (C) $2z$
   (D) $z\cos\theta\sin\theta$
   (E) $z(\cos\theta + \sin\theta)$

17. If the vertices of a triangle are $(u,0)$, $(v,8)$, and $(0,0)$, then the area of the triangle is

   (A) $4|u|$
   (B) $2|v|$
   (C) $|uv|$
   (D) $2|uv|$
   (E) $\dfrac{1}{2}|uv|$

18. What is the difference between the min and max values of the function f defined by $f(x) = 2x^2 + 3x - 8$ on the interval $[-2,5]$?

   (A) 7
   (B) 57.75
   (C) 63.0
   (D) 66.13
   (E) 71.25

19. What is the probability that a prime number is less than 7, given that it is less than 13?

   (A) $\dfrac{1}{3}$

   (B) $\dfrac{2}{5}$

   (C) $\dfrac{1}{2}$

   (D) $\dfrac{3}{5}$

   (E) $\dfrac{3}{4}$

20. The ellipse $4x^2 + 8y^2 = 64$ and the circle $x^2 + y^2 = 9$ intersect at points where the y-coordinate is

(A) $\pm\sqrt{2}$

(B) $\pm\sqrt{5}$

(C) $\pm\sqrt{6}$

(D) $\pm\sqrt{7}$

(E) ± 10.00

21. What value of x makes the three consecutive terms

$x, \dfrac{x}{x+3}, \dfrac{4x}{(x+3)(x+2)}$ the terms of a geometric

sequence?

(A) $-\dfrac{10}{3}$

(B) $-\dfrac{1}{4}$

(C) $\dfrac{2}{3}$

(D) 1

(E) 4

22. A company offers you the use of its computer for a fee. Plan A costs $6 to join and then $9 per hour to use the computer. Plan B costs $25 to join and then $2.25 per hour to use the computer. After how many minutes of use would the cost of Plan A be the same as the cost of Plan B?

(A) 18,052

(B) 173

(C) 169

(D) 165

(E) 157

23. If the probability that the Giants will win the NFC championship is p and if the probability that the Raiders will win the AFC championship is q, what is the probability that only one of these teams will win its respective championship?

(A) pq

(B) $p + q - 2pq$

(C) $|p - q|$

(D) $1 - pq$

(E) $2pq - p - q$

24. What is the sum of the infinite geometric series whose first two terms are 3 and 1?

(A) 1.5

(B) 4.5

(C) 9

(D) 12

(E) infinite

25. Twelve boys try out for the basketball team. In how many ways can 7 of these boys be selected for the team?

   (A) 19
   (B) 32
   (C) 84
   (D) 792
   (E) 95,040

26. If A is the angle formed by the line $2y = 3x + 7$ and the x-axis, then $\angle A$ equals

   (A) −45°
   (B) 0°
   (C) 56°
   (D) 72°
   (E) 215°

27. Suppose $f(x) = \dfrac{1}{x}$. Then $\dfrac{f(x+h) - f(x)}{h} =$

   (A) 1

   (B) $\dfrac{1}{h^2}$

   (C) $\dfrac{x+h}{h}$

   (D) $-\dfrac{1}{x(x+h)}$

   (E) $-\dfrac{x}{x+h}$

28. If $(x-4)^2 + 4(y-3)^2 = 16$ is graphed, the sum of the distances from any point on the curve to the two foci is

   (A) 4
   (B) 8
   (C) 12
   (D) 16
   (E) 32

29. In the equation $x^2 + kx + 54 = 0$, one root is twice the other root. The value(s) of k is (are)

   (A) −5.2
   (B) 15.6
   (C) 22.0
   (D) ±5.2
   (E) ±15.6

30. The remainder obtained when $3x^4 + 7x^3 + 8x^2 - 2x - 3$ is divided by $x + 1$ is

   (A) −3
   (B) 0
   (C) 3
   (D) 5
   (E) 13

31. If $f(x) = e^x$ and $g(x) = f(x) + f^{-1}(x)$, what does $g(2)$ equal?

(A) 5.1
(B) 7.4
(C) 7.5
(D) 8.1
(E) 8.3

32. If $x_0 = 3$ and $x_{n+1} = \sqrt{4 + x_n}$, then $x_3 =$

(A) 2.65
(B) 2.58
(C) 2.56
(D) 2.55
(E) 2.54

33. For what values of k does the graph of

$$\frac{(x - 2k)^2}{1} - \frac{(y - 3k)^2}{3} = 1 \text{ pass through the origin?}$$

(A) only 0
(B) only 1
(C) ±1
(D) $\pm\sqrt{5}$
(E) no value

34. If $\dfrac{1 - \cos\theta}{\sin\theta} = \dfrac{\sqrt{3}}{3}$, then $\theta =$

(A) 15°
(B) 30°
(C) 45°
(D) 60°
(E) 75°

35. If $x^2 + 3x + 2 < 0$ and $f(x) = x^2 - 3x + 2$, then

(A) $0 < f(x) < 6$
(B) $f(x) \geq \dfrac{3}{2}$
(C) $f(x) > 12$
(D) $f(x) > 0$
(E) $6 < f(x) < 12$

36. If $f(x) = |x| + [x]$, where $[x]$ is the greatest integer less than or equal to x, the value of $f(-2.5) + f(1.5)$ is

(A) –2
(B) 1
(C) 1.5
(D) 2
(E) 3

37. If $(\sec x)(\tan x) < 0$, which of the following must be true?

   I. $\tan x < 0$
  II. $\csc x \cot x < 0$
 III. x is in the third or fourth quadrant

(A) I only
(B) II only
(C) III only
(D) II and III
(E) I and II

38. At the end of a meeting all participants shook hands with each other. Twenty-eight handshakes were exchanged. How many people were at the meeting?

(A) 7
(B) 8
(C) 14
(D) 28
(E) 56

39. Suppose the graph of $f(x) = 2x^2$ is translated 3 units down and 2 units right. If the resulting graph represents the graph of $g(x)$, what is the value of $g(-1.2)$?

(A) -1.72
(B) -0.12
(C) 2.88
(D) 17.48
(E) 37.28

x	-5	-3	-1	1
y	0	4	-3	0

40. Four points on the graph of a polynomial P are shown in the table above. If P is a polynomial of degree 3, then $P(x)$ could equal

(A) $a_1(x-5)(x-2)(x+1)$
(B) $a_2(x-5)(x+2)(x+1)$
(C) $a_3(x+5)(x-2)(x-1)$
(D) $a_4(x+5)(x+2)(x-1)$
(E) $a_5(x+5)(x+2)(x+1)$

41. If $f(x) = ax + b$, which of the following make(s) $f(x) = f^{-1}(x)$?

   I. $a = -1$, $b =$ any real number
  II. $a = 1$, $b = 0$
 III. $a =$ any real number, $b = 0$

(A) only I
(B) only II
(C) only III
(D) only I and II
(E) only I and III

42. In the figure above, $\angle A = 110°$, $a = \sqrt{6}$, and $b = 2$. What is the value of $\angle C$?

(A) 50°
(B) 25°
(C) 20°
(D) 15°
(E) 10°

43. If vector $\vec{v} = \left(1, \sqrt{3}\right)$ and vector $\vec{u} = (3, -2)$, find the value of $\left| 3\vec{v} - \vec{u} \right|$.

(A) 5.4
(B) 6
(C) 7
(D) 7.2
(E) 52

44. If $f(x) = \sqrt{x^2 - 1}$ and $g(x) = \dfrac{10}{x + 2}$, then $g(f(3)) =$

(A) 0.2
(B) 1.7
(C) 2.1
(D) 3.5
(E) 8.7

USE THIS SPACE FOR SCRATCH WORK

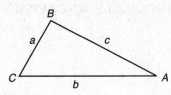

45. In $\triangle ABC$ above, $a = 2x$, $b = 3x + 2$, $c = \sqrt{12}$, and $\angle C = 60°$. Find x.

(A) 0.50
(B) 0.64
(C) 0.77
(D) 1.64
(E) 1.78

46. If $\log_a 5 = x$ and $\log_a 7 = y$, then $\log_a \sqrt{1.4} =$

(A) $\dfrac{1}{2}xy$

(B) $\dfrac{1}{2}(x - y)$

(C) $\dfrac{1}{2}(x + y)$

(D) $\dfrac{1}{2}(y - x)$

(E) $\dfrac{y}{2x}$

47. If $f(x) = 3x^2 + 4x + 5$, what must the value of k equal so that the graph of $f(x - k)$ will be symmetric about the y-axis?

(A) -4

(B) $-\dfrac{4}{3}$

(C) $-\dfrac{2}{3}$

(D) $\dfrac{2}{3}$

(E) $\dfrac{4}{3}$

48. If $f(x) = \cos x$ and $g(x) = 2x + 1$, which of the following are even functions?

   I. $f(x) \cdot g(x)$
  II. $f(g(x))$
 III. $g(f(x))$

  (A) only I
  (B) only II
  (C) only III
  (D) only I and II
  (E) only II and III

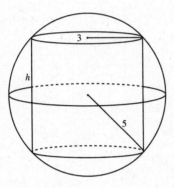

49. A cylinder whose base radius is 3 is inscribed in a sphere of radius 5. What is the difference between the volume of the sphere and the volume of the cylinder?

  (A) 88
  (B) 297
  (C) 354
  (D) 448
  (E) 1,345

50. What does $\dfrac{3 - 2x + x^3}{7 + 2x + x^2 - 5x^3}$ get closer to as x gets smaller?

  (A) $-\infty$
  (B) -0.2
  (C) 0.2
  (D) $\dfrac{3}{7}$
  (E) ∞

STOP

If there is still time remaining, you may review your answers.

ANSWER KEY
Practice Test 1

1. **C**	14. **D**	27. **D**	40. **D**
2. **A**	15. **C**	28. **B**	41. **D**
3. **A**	16. **B**	29. **E**	42. **C**
4. **C**	17. **A**	30. **C**	43. **D**
5. **D**	18. **D**	31. **D**	44. **C**
6. **B**	19. **D**	32. **C**	45. **B**
7. **B**	20. **D**	33. **C**	46. **D**
8. **B**	21. **A**	34. **D**	47. **D**
9. **E**	22. **C**	35. **E**	48. **C**
10. **D**	23. **B**	36. **D**	49. **B**
11. **C**	24. **B**	37. **C**	50. **B**
12. **C**	25. **D**	38. **B**	
13. **E**	26. **C**	39. **D**	

ANSWERS EXPLAINED

The following explanations are keyed to the review portions of this book. The number in brackets after each explanation indicates the appropriate chapter in the Review of Major Topics (Part 2).

An asterisk appears next to those solutions for which a graphing calculator is necessary or helpful.

1. **(C)** Solve for y. $y = \frac{4}{3}x - 4$. Slope $= \frac{4}{3}$. The slope of a perpendicular line is the opposite reciprocal, $-\frac{3}{4}$. [2]

2. **(A)** Range = largest value − smallest value = $18 - 8 = 10$. [21]

3. **(A)** Vertical asymptotes occur where the denominator is zero but the numerator is not. The denominator, $x^2 - 49$, factors into $(x + 7)(x - 7)$. Since both the numerator and the denominator are zero when $x = 7$, a vertical asymptote occurs only at $x = -7$. [9]

* 4. **(C)** Enter the function f into Y_1 and the function g into Y_2. Evaluate $Y_1(Y_2(2))$ to get the correct answer, Choice C.

   An alternative solution is to evaluate $g(2) = 5$ and $f(5) = \sqrt{13}$, and either use your calculator to evaluate $\sqrt{13}$ or observe that $3 < \sqrt{13} < 4$, indicating that 3.61 is the only feasible answer choice. [1]

* 5. **(D)** $4^{x+1} = (2^2)^{x+1} = 2^{2x+2}$ and $8^x = (2^3)^x = 2^{3x}$. Therefore, $2^{2x+2} = 2^{3x}$. It follows that $2x + 2 = 3x$, and $x = 2$. [8]

* 6. **(B)** Complete the square to get $x^2 + (y - 5)^2 = 61$.
   Radius $= \sqrt{61}$. $C = 2\pi r = 2\pi\sqrt{61} \approx 49$. [15]

* 7. **(B)** Whether or not students in the group have siblings are independent events. The probability that each of the first four is not an only child is $(0.75)^4$. The probability that the fifth student is an only child is 0.25, so the probability of seeing the first four children with siblings and the fifth as an only child is $(0.75)^4(0.25) \approx 0.08$. [22]

8. **(B)** Regardless of what is substituted for x, $f(x)$ still equals 2.

   Alternative Solution: $f(x + 2)$ causes the graph of $f(x)$ to be shifted 2 units to the left. Since $f(x) = 2$ for all x, $f(x + 2)$ will also equal 2 for all x. [1]

* 9. **(E)** Enter the formula for the volume of a sphere $(4/3)\pi x^3$ (in the reference list of formulas) into Y_1. Return to the home screen, and enter $Y_1(5) - Y_1(2)$ to get the correct answer, Choice E.

   An alternative solution is to evaluate $V = \frac{4}{3}\pi(5^3 - 2^3) = \frac{4}{3}\pi(117)$ directly. [15]

* 10. **(D)** Since $12 = 2^2 \times 3$, $\log_b 12 = \log_b 2^2 + \log_b 3 = 2\log_b 2 + \log_b 3 =$ $2(0.4307) + 0.6826 = 1.5440$. [8]

11. **(C)** $\sin A = \dfrac{8}{10} = \dfrac{4}{5}$. [6]

12. **(C)** Since the $\dfrac{1}{2}$ power is the square root, $x = \dfrac{1}{2}$ because the square root of 16 is 4.

Since $5^4 = 625$, $x + y = 4$, which means that $y = 4 - \dfrac{1}{2} = \dfrac{7}{2}$. [8]

13. **(E)** $t = \dfrac{y}{2}$. Eliminate the parameter and get $x = \dfrac{y^2}{4} + 1 = x$ or $y^2 = 4x - 4$. [10]

14. **(D)** $f(1) = 0$ and $f(2) = 0$ imply that $x - 1$ and $x - 2$ are factors of $f(x)$. Their product, $x^2 - 3x + 2$, is also a factor. [4]

15. **(C)** In order for the numerator to be a real number, $25 - x^2 \geq 0$. Therefore, $-5 \leq x \leq 5$. In order for the denominator to be a non-zero real number, $x^2 - 9 > 0$. Therefore, $x < -3$ or $x > 3$. Since both conditions must be met, $-5 \leq x < -3$ and $3 < x \leq 5$. This is the interval $[-5,-3) \cup (3,5]$. [1]

16. **(B)** $a^2 = z^2 \cos^2 \theta$ and $b^2 = z^2 \sin^2 \theta$, so $a^2 + b^2 = z^2(\cos^2 \theta + \sin^2 \theta) = z^2$ because $\cos^2 \theta + \sin^2 \theta = 1$. Since $\sqrt{z^2} = z$ when $z > 0$, the correct answer choice is Choice B. [7]

17. **(A)** Sketch a graph of the three vertices. The base is $|u|$ and the altitude is 8. Therefore, the area is $4|u|$. [15]

* 18. **(D)** Enter the function into Y_1 in the graphing calculator, with XMIN = -2 and XMAX = 5. In order to see the entire graph, you will have to set YMIN = -10 and YMAX = 60. Use CALC (2nd TRACE) to find the min value -9.13 and the max value 57. The difference is $57 - (-9.13) = 66.13$. [4]

19. **(D)** There are 5 prime numbers less than 13: 2, 3, 5, 7, 11. Three of these are less than 7, so the correct probability is $\dfrac{3}{5}$. [22]

* 20. **(D)** Substituting for x^2 and solving for y gives $4(9 - y^2) + 8y^2 = 64$. $4y^2 = 28$, and so $y^2 = 7$ and $y = \pm\sqrt{7}$. [13]

21. **(A)** In order for these three consecutive terms to be in a geometric sequence, the ratio of the second term to the first term must equal the ratio of the third term to the second term.

In other words, $\dfrac{\frac{x}{x+3}}{x} = \dfrac{\frac{4x}{(x+3)(x+2)}}{\frac{x}{x+3}}$. Simplifying each side yields $\dfrac{1}{x+3} = \dfrac{4}{x+2}$,

and cross-multiplying gives the equation $4x + 12 = x + 2$. Therefore, $x = -\dfrac{10}{3}$. [19]

* 22. **(C)** Graph the cost of Plan A, $y = 6 + 9x$, and the cost of Plan B, $y = 25 + 2.25x$, in a window $x\varepsilon[0,10]$ and $y\varepsilon[0,50]$. Use CALC/intersect to find the x-coordinate of the point of intersection at 2.8148 hours, the "break-even" point. Multiply by 60 to convert this time to the correct answer choice.

An alternative solution is to solve the equation $6 + 9x = 25 + 2.25x$ and multiply the solution by 60 to get the answer of about 169 minutes. [2]

23. **(B)** The probability that both teams will win is pq. The probability that both will lose is $(1 - p)(1 - q)$. The probability that only one will win is:

$$1 - [pq + (1 - p)(1 - q)] = 1 - (pq + 1 - p - q + pq) = p + q - 2pq$$

Alternative Solution: The probability that the Giants will win and the Raiders will lose is $p(1 - q)$. The probability that the Raiders will win and the Giants will lose is $q(1 - p)$. Therefore, the probability that either one of these results will occur is $p(1 - q) + q(1 - p) = p + q - 2pq$. [22]

24. **(B)** The common ratio r of this geometric sequence is $\dfrac{1}{3}$, and the first term is 3.

The formula for the sum of an infinite series is $S = \dfrac{a}{1-r}$, where a is the first term and r is the common ratio. Therefore, $S = \dfrac{3}{1-\frac{1}{3}} = \dfrac{3}{\frac{2}{3}} = 4.5$. [19]

* 25. **(D)** Seven people are being chosen from 12. The number of ways this can be done is

$_{12}C_7 = \dfrac{12!}{7!5!} = 792$. [16]

* 26. **(C)** Solve for y: $y = \dfrac{3}{2}x + \dfrac{7}{2}$. Slope $= \dfrac{\Delta y}{\Delta x} = \dfrac{3}{2}$. Tan A also equals $\dfrac{\Delta y}{\Delta x}$. Therefore,

$\tan A = \dfrac{3}{2}$. $\text{Tan}^{-1}\left(\dfrac{3}{2}\right) = \angle A \approx 56°$. [6]

27. **(D)** $\dfrac{f(x+h)-f(x)}{h} = \dfrac{\frac{1}{x+h} - \frac{1}{x}}{h} = \dfrac{\frac{x-(x+h)}{x(x+h)}}{h} = \dfrac{-h}{hx(x+h)} = \dfrac{-1}{x(x+h)}$. [1]

28. **(B)** Divide the equation through by 16 to get $\dfrac{(x-4)^2}{16} + \dfrac{(y-3)^2}{4} = 1$. This is the equation of an ellipse with $a^2 = 16$. The sum of the distances to the foci is $2a = 8$. [13]

* 29. **(E)** If the roots are r and $2r$, their sum is $-\dfrac{b}{a} = 3r = -\dfrac{k}{1}$ and their product is

$\dfrac{c}{a} = 2r^2 = \dfrac{54}{1}$. Therefore, $r = \pm\sqrt{27}$ and $k = \pm 3\sqrt{27} \approx \pm 15.6$.

Alternative Solution: If the roots are r and $2r$, $(x-r)(x-2r) = 0$. Multiply to obtain $x^2 - 3r + 2r^2 = 0$, which represents $x^2 + kx + 54 = 0$. Thus, $-3r = k$ and $2r^2 = 54$.

Since $r = -\dfrac{k}{3}$, then $2\left(-\dfrac{k}{3}\right) = 54$ and $k = \pm 3\sqrt{27} \approx \pm 15.6$. [3]

30. **(C)** According to the Remainder Theorem, the remainder, upon division by $x + 1$, is the value of the dividend when -1 is substituted for x:

$$3(-1)^4 + 7(-1)^3 + 8(-1)^2 - 2(-1) - 3 = 3 \quad [4]$$

* 31. **(D)** The inverse of $f(x) = e^x$ is $f^{-1}(x) = \ln x$. $g(2) = e^2 + \ln 2 \approx 8.1$. [8]

* 32. **(C)** This is a recursively defined sequence. Press 3 ENTER on your calculator.

Then enter $\sqrt{4 + \text{Ans}}$ and press ENTER 3 times to get $x_3 \approx 2.56$. [19]

33. **(C)** If the graph passes through the origin, $x = 0$ and $y = 0$, then $\dfrac{4k^2}{1} - \dfrac{9k^2}{3} = 1$. $k^2 = 1$, and so $k = \pm 1$. [13]

* 34. **(D)** Graph $y = \dfrac{1 - \cos x}{\sin x}$ and $y = \dfrac{\sqrt{3}}{3}$ using Ztrig in degree mode. Find the point of

intersection with CALC/intersect to arrive at the correct answer, Choice D.

An alternative solution uses the identities $\tan\dfrac{\theta}{2} = \dfrac{1 - \cos x}{\sin x}$ and $\tan 30° = \dfrac{\sqrt{3}}{3}$ to

deduce $\dfrac{\theta}{2} = 30°$, so $\theta = 60°$. [7]

* 35. **(E)** The problem is asking for the range of $f(x)$ values for values of x that satisfy the inequality. First graph the inequality in Y_1, starting with the standard window and zooming in until the x-values for the portion of the graph that falls below the x-axis can be identified as the interval $(-2,-1)$. Then enter the formula for $f(x)$ in Y_2. Although it can be done graphically, the simplest way to find the range of values of $f(x)$ that correspond to $x\epsilon(-2,-1)$ is to use the TABLE function. Deselect Y_1 and enter TBLSET and set TblStart to -2, ΔTbl $= 0.1$, and Indpnt and Depend to Auto. Then enter TABLE and observe that the Y_2 values range from 12 to 6 as x ranges from -2 to -1, yielding the correct answer, Choice E.

An alternative solution is to solve the inequality algebraically by solving the associated equation $x^2 + 3x + 2 = 0$ and testing points. The left side of the equation factors as $(x + 2)(x + 1)$, and the Zero Product Property implies that $x = -2$ or $x = -1$. Points inside the interval $(-2,-1)$ satisfy the inequality, while those outside it do not. Since the graph of $f(x)$ is a parabola and $f(-2) = 12$ and $f(-1) = 6$, $f(x)$ takes the range of values between 12 and 6. [5]

*36.   **(D)** The greatest integer less than or equal to x is int(x) on the graphing calculator. Enter abs(x) + int(x) into Y_1. Return to the home screen, and enter $Y_1(-2.5) + Y_1(1.5)$ to get the correct answer, Choice D.

An alternative solution evaluates $|-2.5| + [-2.5] + |1.5| + [1.5]$ without the aid of a calculator. Of these 4 values, only $[-2.5]$ is tricky since $[-2.5] = -3$, not -2. Thus, $|-2.5| + [-2.5] + |1.5| + [1.5] = 2.5 - 3 + 1.5 + 1 = 2$. [11]

37.   **(C)** Set up the following table.

	Q1	Q2	Q3	Q4
$\sec x$	+	–	–	+
$\tan x$	+	–	+	–
$\cot x$	+	–	+	–
$\csc x$	+	+	–	–

The product $\sec x \tan x$ is negative only when its factors have different signs, so III is the only true statement. [7]

38.   **(B)** $\begin{pmatrix} x \text{ people} \\ 2 \end{pmatrix} = 28$. $\dfrac{x(x-1)}{2 \cdot 1} = 28$. $x^2 - x = 56$. $x = 8$. [16]

*39.   **(D)** Since the function g is f translated 3 down and 2 right, $g(x) = f(x - 2) - 3$. Therefore, $g(-1.2) = f(-3.2) - 3 = 2(-3.2)^2 - 3 = 17.48$. [12]

40.   **(D)** Since -5 and 1 are both zeros, $(x + 5)$ and $(x - 1)$ are factors of $P(x)$. Since $P(x)$ changes sign between $x = -3$ and $x = -1$, there is a zero between these two values. Choice D is the only one that meets all three criteria. [4]

*41.   **(D)** The graph of f must be symmetric about the line $y = x$. In other words, interchanging x and y must leave the graph unchanged. In I, $x = -y + b$, which is equivalent to $y = -x + b$, which is symmetric about $y = x$. In II, $x = y$. In III, $x = ay$, or $y = \dfrac{x}{a}$. [2]

*42.   **(C)** Law of Sines:

$$\frac{\sin 110°}{\sqrt{6}} = \frac{\sin B}{2}$$

$$\sin B = \frac{2 \sin 110°}{\sqrt{6}} \approx 0.7673$$

$$\text{Sin}^{-1}(0.7673) = \angle B = 50°$$

Therefore, $\angle C = 180° - 110° - 50° = 20°$. [6]

*43.   **(D)** $\left| 3\vec{v} - \vec{u} \right| = \left| (3, 3\sqrt{3}) - (3, -2) \right| = \left| (0, 3\sqrt{3} + 2) \right| = \sqrt{0^2 + (3\sqrt{3} + 2)^2} = 3\sqrt{3} + 2 \approx 7.2.$

[20]

*44.   **(C)** $f(3) = \sqrt{8}$ and $g(\sqrt{8}) = \dfrac{10}{\sqrt{8} + 2} \approx 2.1$. [1]

✱ **45.** **(B)** Law of Cosines:

$$12 = (2x)^2 + (3x+2)^2 - 2 \cdot 2x \cdot (3x+2)\cos 60°$$

$$12 = 4x^2 + 9x^2 + 12x + 4 - (12x^2 + 8x) \cdot \frac{1}{2}$$

$$7x^2 + 8x - 8 = 0$$

Use the quadratic formula to get $x = \pm 0.64$.

Since a side of a triangle must be positive, x can only equal 0.64. [6]

46. **(D)** $\log_a \sqrt{1.4} = \log_a \left(\frac{7}{5}\right)^{1/2} = \frac{1}{2}\left(\log_a 7 - \log_a 5\right) = \frac{1}{2}(y - x)$. [8]

✱ **47.** **(D)** Graph $y = 3x^2 + 4x + 5$ in the standard window, and observe that the graph must be moved slightly to the right to be symmetric to the y-axis. Therefore, k must be positive. Use CALC/minimum to find the vertex of the parabola and observe that its

x-coordinate is $-0.66666\ldots$. If the function entered into Y_1, set $Y_2 = Y_1\left(x - \frac{2}{3}\right)$

and graph Y_2 to verify this answer. [12]

✱ **48.** **(C)** Use ZTrig to plot the graphs of $y = (\cos x) \cdot (2x + 1)$, $y = \cos(2x + 1)$, and $y = 2(\cos x) + 1$ to see that only the third graph is symmetric about the y-axis and thus represents an even function.

An alternative solution is to use your knowledge of transformations. Although f is an even function, g is not; therefore, (I) $f \cdot g$ is not even. Also, $f(g(x)) = \cos(2x + 1)$, which is a cosine curve shifted less than π to the left. Thus, $f(g(x))$ (II) is not even. However, $g(f(x)) = 2\cos x + 1$ is a cosine curve with period 2π, amplitude 2, shifted 1 unit up. Thus, $g(f(x))$ (III) is even. [1]

✱ **49.** **(B)** The height of the cylinder is 8.

The volume of the sphere is $\frac{4}{3}\pi r^3 = \frac{4}{3}\pi(125) = \frac{500\pi}{3}$.

The volume of the cylinder is $\pi r^2 h = \pi(9)8$.

The difference is $\frac{500}{3}\pi - 72\pi \approx 523.6 - 226.7$

$$\approx 297 \; [15]$$

50. **(B)** The x^3 terms "dominate" both the numerator and the denominator for small values of x, such as -10^{100}. For such values, this rational expression gets closer to

$\dfrac{x^3}{-5x^3} = -\dfrac{1}{5} = -0.2$. [9]

Self-Evaluation Chart for Practice Test 1

Subject Area	Questions and Review Section						Right	Number Wrong	Omitted
Algebra and Functions (24 questions)	1 2	3 9	4 1	5 8	8 1	10 8	____	____	____
	12 8	13 10	14 4	15 1	18 4	22 2	____	____	____
	27 1	29 3	30 4	31 8	35 5	36 11	____	____	____
	40 4	41 2	44 1	46 8	48 1	50 9	____	____	____
Trigonometry (7 questions)	11 6	16 7	26 6	34 7	37 7	42 6	____	____	____
	45 6						____	____	____
Coordinate and Three-Dimensional Geometry (9 questions)	6 15	9 15	17 15	20 13	28 13	33 13	____	____	____
	39 12	47 12	49 15				____	____	____
Number and Operations (6 questions)	21 19	24 19	25 16	32 19	38 16	43 20	____	____	____
Data Analysis, Statistics, and Probability (4 questions)	2 21	7 22	19 22	23 22			____	____	____
TOTALS							____	____	____

Evaluate Your Performance on Practice Test 1	
Rating	**Number Right**
Excellent	41–50
Very good	33–40
Above average	25–32
Average	15–24
Below average	Below 15

CALCULATING YOUR SCORE

Raw score R = number right $- \dfrac{1}{4}$ (number wrong), rounded = _____

Approximate scaled score $S = 800 - 10(44 - R) =$ _____

If $R \geq 44$, $S = 800$.

ANSWER SHEET
Practice Test 2

1. Ⓐ Ⓑ Ⓒ Ⓓ Ⓔ
2. Ⓐ Ⓑ Ⓒ Ⓓ Ⓔ
3. Ⓐ Ⓑ Ⓒ Ⓓ Ⓔ
4. Ⓐ Ⓑ Ⓒ Ⓓ Ⓔ
5. Ⓐ Ⓑ Ⓒ Ⓓ Ⓔ
6. Ⓐ Ⓑ Ⓒ Ⓓ Ⓔ
7. Ⓐ Ⓑ Ⓒ Ⓓ Ⓔ
8. Ⓐ Ⓑ Ⓒ Ⓓ Ⓔ
9. Ⓐ Ⓑ Ⓒ Ⓓ Ⓔ
10. Ⓐ Ⓑ Ⓒ Ⓓ Ⓔ
11. Ⓐ Ⓑ Ⓒ Ⓓ Ⓔ
12. Ⓐ Ⓑ Ⓒ Ⓓ Ⓔ
13. Ⓐ Ⓑ Ⓒ Ⓓ Ⓔ

14. Ⓐ Ⓑ Ⓒ Ⓓ Ⓔ
15. Ⓐ Ⓑ Ⓒ Ⓓ Ⓔ
16. Ⓐ Ⓑ Ⓒ Ⓓ Ⓔ
17. Ⓐ Ⓑ Ⓒ Ⓓ Ⓔ
18. Ⓐ Ⓑ Ⓒ Ⓓ Ⓔ
19. Ⓐ Ⓑ Ⓒ Ⓓ Ⓔ
20. Ⓐ Ⓑ Ⓒ Ⓓ Ⓔ
21. Ⓐ Ⓑ Ⓒ Ⓓ Ⓔ
22. Ⓐ Ⓑ Ⓒ Ⓓ Ⓔ
23. Ⓐ Ⓑ Ⓒ Ⓓ Ⓔ
24. Ⓐ Ⓑ Ⓒ Ⓓ Ⓔ
25. Ⓐ Ⓑ Ⓒ Ⓓ Ⓔ
26. Ⓐ Ⓑ Ⓒ Ⓓ Ⓔ

27. Ⓐ Ⓑ Ⓒ Ⓓ Ⓔ
28. Ⓐ Ⓑ Ⓒ Ⓓ Ⓔ
29. Ⓐ Ⓑ Ⓒ Ⓓ Ⓔ
30. Ⓐ Ⓑ Ⓒ Ⓓ Ⓔ
31. Ⓐ Ⓑ Ⓒ Ⓓ Ⓔ
32. Ⓐ Ⓑ Ⓒ Ⓓ Ⓔ
33. Ⓐ Ⓑ Ⓒ Ⓓ Ⓔ
34. Ⓐ Ⓑ Ⓒ Ⓓ Ⓔ
35. Ⓐ Ⓑ Ⓒ Ⓓ Ⓔ
36. Ⓐ Ⓑ Ⓒ Ⓓ Ⓔ
37. Ⓐ Ⓑ Ⓒ Ⓓ Ⓔ
38. Ⓐ Ⓑ Ⓒ Ⓓ Ⓔ
39. Ⓐ Ⓑ Ⓒ Ⓓ Ⓔ

40. Ⓐ Ⓑ Ⓒ Ⓓ Ⓔ
41. Ⓐ Ⓑ Ⓒ Ⓓ Ⓔ
42. Ⓐ Ⓑ Ⓒ Ⓓ Ⓔ
43. Ⓐ Ⓑ Ⓒ Ⓓ Ⓔ
44. Ⓐ Ⓑ Ⓒ Ⓓ Ⓔ
45. Ⓐ Ⓑ Ⓒ Ⓓ Ⓔ
46. Ⓐ Ⓑ Ⓒ Ⓓ Ⓔ
47. Ⓐ Ⓑ Ⓒ Ⓓ Ⓔ
48. Ⓐ Ⓑ Ⓒ Ⓓ Ⓔ
49. Ⓐ Ⓑ Ⓒ Ⓓ Ⓔ
50. Ⓐ Ⓑ Ⓒ Ⓓ Ⓔ

Practice Test 2

Tear out the preceding answer sheet. Decide which is the best choice by rounding your answer when appropriate. Blacken the corresponding space on the answer sheet. When finished, check your answers with those at the end of the test. For questions that you got wrong, note the chapters containing the material that you must review. Also, if you do not fully understand how you arrived at some of the correct answers, you should review the appropriate chapters. Finally, fill out the Self-Evaluation Chart on page 251 in order to pinpoint the topics that gave you the most difficulty.

50 questions: 1 hour

Directions: Decide which answer choice is best. If the exact numerical value is not one of the answer choices, select the closest approximation. Fill in the circle on the answer sheet that corresponds to your choice.

Notes:
(1) You will need to use a scientific or graphing calculator to answer some of the questions.
(2) You will have to decide whether to put your calculator in degree or radian mode for some problems.
(3) All figures that accompany problems are plane figures unless otherwise stated. Figures are drawn as accurately as possible to provide useful information for solving the problem, except when it is stated in a particular problem that the figure is not drawn to scale.
(4) Unless otherwise indicated, the domain of a function is the set of all real numbers for which the functional value is also a real number.

Reference Information. The following formulas are provided for your information.

Volume of a right circular cone with radius r and height h: $V = \dfrac{1}{3}\pi r^2 h$

Volume of a sphere of radius r: $V = \dfrac{4}{3}\pi r^3$

Surface area of a sphere of radius r: $S = 4\pi r^2$

Volume of a pyramid of base area B and height h: $V = \dfrac{1}{3}Bh$

1. If $f(x) = \dfrac{x-2}{x^2-4}$, for what value(s) of x does the graph of $f(x)$ have a vertical asymptote?

   (A) $-2, 0,$ and 2
   (B) -2 and 2
   (C) 2
   (D) 0
   (E) -2

2. What is the distance between the points with coordinates $(-3,4,1)$ and $(2,7,-4)$?

   (A) 5.24
   (B) 7.68
   (C) 11.45
   (D) 13.00
   (E) 19.26

3. $\mathrm{Log}(a^2 - b^2) =$

   (A) $\log a^2 - \log b^2$

   (B) $\log \dfrac{a^2}{b^2}$

   (C) $\log \dfrac{a+b}{a-b}$

   (D) $2 \cdot \log a - 2 \cdot \log b$

   (E) $\log(a+b) + \log(a-b)$

4. The domain of the function $f(x) = 4 - \sqrt{x^2 - 9}$ is

   (A) $x < -3$
   (B) $x > 0$
   (C) $x > 3$
   (D) $x \le -3$ or $x \ge 3$
   (E) all real numbers

5. If the graph of $x + 2y + 3 = 0$ is perpendicular to the graph of $ax + 3y + 2 = 0$, then a equals

   (A) -6

   (B) $-\dfrac{3}{2}$

   (C) $\dfrac{2}{3}$

   (D) $\dfrac{3}{2}$

   (E) 6

6. The maximum value of $6 \sin x \cos x$ is

(A) $\dfrac{1}{3}$

(B) 1
(C) 2.6
(D) 3
(E) 6

7. If $f(r,\theta) = r\cos\theta$, then $f(2,3) =$

(A) -3.00
(B) -1.98
(C) 0.10
(D) 1.25
(E) 2.00

8. If 5 and -1 are both zeros of the polynomial $P(x)$, then a factor of $P(x)$ is

(A) $x^2 - 5$
(B) $x^2 - 4x + 5$
(C) $x^2 + 4x - 5$
(D) $x^2 + 5$
(E) $x^2 - 4x - 5$

9. $i^{14} + i^{15} + i^{16} + i^{17} =$

(A) 0
(B) 1
(C) $2i$
(D) $1 - i$
(E) $2 + 2i$

10. If the length of an arc of a circle is 4π, and the area of the sector formed by that arc is 8π, what is the radius of the circle?

(A) 1
(B) 2
(C) 4
(D) 8
(E) 12

11. The third term of an arithmetic sequence is 15, and the seventh term is 23. What is the first term?

(A) 1
(B) 6
(C) 9
(D) 11
(E) 13

12. A particular sphere has the property that its surface area has the same numerical value as its volume. What is the length of the radius of this sphere?

(A) 1
(B) 2
(C) 3
(D) 4
(E) 6

13. What value does $\dfrac{x^4 - 3x^3 + 6x + 1}{x^5 - 3x^2 + 100}$ get closer to as x gets larger?

(A) 0

(B) $\dfrac{1}{100}$

(C) 1
(D) 100
(E) ∞

14. The pendulum on a clock swings through an angle of 1 radian, and the tip sweeps out an arc of 12 inches. How long is the pendulum?

(A) 3.8 inches
(B) 6 inches
(C) 7.6 inches
(D) 12 inches
(E) 35 inches

15. What is the domain of the function

$f(x) = 4 - \sqrt{3x^3 - 7}$?

(A) $x \geq 1.33$
(B) $x \geq 1.53$
(C) $x \geq 2.33$
(D) $x \leq -1.33$ or $x \geq 1.33$
(E) $x \leq -2.33$ or $x \geq 2.33$

16. If $x + y = 90°$, which of the following must be true?

(A) $\cos x = \cos y$
(B) $\sin x = -\sin y$
(C) $\tan x = \cot y$
(D) $\sin x + \cos y = 1$
(E) $\tan x + \cot y = 1$

17. The graph of the equation $y = x^3 + 5x + 1$

   (A) does not intersect the x-axis
   (B) intersects the x-axis at one and only one point
   (C) intersects the x-axis at exactly three points
   (D) intersects the x-axis at more than three points
   (E) intersects the x-axis at exactly two points

18. The length of the radius of the sphere
$x^2 + y^2 + z^2 + 2x - 4y = 10$ is

   (A) 3.16
   (B) 3.38
   (C) 3.46
   (D) 3.74
   (E) 3.87

```
WINDOW
 Xmin=-10
 Xmax=5
 Xscl=1
 Ymin=-250
 Ymax=250
 Yscl=50
 Xres=1
 ΔX=.05681818181818
 TraceStep=.11363636363636
```

19. The graph of $y = x^4 + 11x^3 + 9x^2 - 97x + c$ is shown above with the window shown below it. Which of the following values could be c?

   (A) –2,820
   (B) –80
   (C) 80
   (D) 250
   (E) 2,820

20. Which of the following is the solution set for
$x(x - 3)(x + 2) > 0$?

   (A) $x < -2$
   (B) $-2 < x < 3$
   (C) $-2 < x < 3$ or $x > 3$
   (D) $x < -2$ or $0 < x < 3$
   (E) $-2 < x < 0$ or $x > 3$

21. Which of the following is the equation of the circle that has its center at the origin and is tangent to the line with equation $3x - 4y = 10$?

   (A) $x^2 + y^2 = 2$
   (B) $x^2 + y^2 = 3$
   (C) $x^2 + y^2 = 4$
   (D) $x^2 + y^2 = 5$
   (E) $x^2 + y^2 = 10$

22. If $f(x) = 3 - 2x + x^2$, then $\dfrac{f(x+t) - f(x)}{t} =$

   (A) $t^2 + 2xt - 2t$
   (B) $x^2t^2 - 2xt + 3$
   (C) $t + 2x - 2$
   (D) $2x - 2$
   (E) none of the above

23. If $f(x) = x^3$ and $g(x) = x^2 + 1$, which of the following is an odd function (are odd functions)?

   I. $f(x) \cdot g(x)$
   II. $f(g(x))$
   III. $g(f(x))$

   (A) only I
   (B) only II
   (C) only III
   (D) only II and III
   (E) I, II, and III

24. In how many ways can a committee of four be selected from nine men so as to always include a particular man?

   (A) 48
   (B) 56
   (C) 70
   (D) 84
   (E) 126

PRACTICE TEST 2

25. The figure above shows a portion of the graph of
$y = 2^x$. What is the sum of the areas of the three
inscribed rectangles shown?

(A) 14
(B) 28
(C) 128
(D) 256
(E) 384

26. If the mean of the set of data 1, 2, 3, 1, 2, 5, x is $3.\overline{27}$,
what is the value of x?

(A) −10.7
(B) 2.5
(C) 5.6
(D) 7.4
(E) 8.9

27. In $\triangle JKL$, $\sin L = \dfrac{1}{3}$, $\sin J = \dfrac{3}{5}$, and $JK = \sqrt{5}$ inches.
The length of KL, in inches, is

(A) 1.7
(B) 3.0
(C) 3.5
(D) 3.9
(E) 4.0

28. Matrix X has r rows and c columns, and matrix Y has c rows and d columns, where r, c, and d are different. Which of the following statements must be true?

   I. The product YX exists.
  II. The product XY exists and has r rows and d columns.
 III. The product XY exists and has c rows and c columns.

(A) I only
(B) II only
(C) III only
(D) I and II
(E) I and III

29. For what value(s) of x is $x^2 + 5 > 1 - 5x$?

(A) $x < -4$
(B) $-4 < x < 0$
(C) $-4 < x < -1$
(D) $x < -4$ or $x > -1$
(E) $1 < x < 4$

30. If $f(x) = x - 7$ and $g(x) = \sqrt{x}$, what is the domain of $g \circ f$?

(A) $x \le 0$
(B) $x \ge -7$
(C) $x \ge 0$
(D) $x \ge 7$
(E) all real numbers

31. In $\triangle ABC$, $a = 1$, $b = 4$, and $\angle C = 30°$. The length of c is

(A) 4.6
(B) 3.6
(C) 3.2
(D) 2.9
(E) 2.3

32. The solution set of $3x + 4y < 0$ lies in which quadrants?

(A) I only
(B) I and II
(C) I, II, and III
(D) II, III, and IV
(E) I, II, III, and IV

PRACTICE TEST 2

33. What is the range of the function defined by

$$f(x) = \frac{5}{x+1} - 4?$$

(A) All real numbers
(B) All real numbers except –4

(C) All real numbers except $-\dfrac{5}{4}$

(D) All real numbers except 0
(E) All real numbers between –4 and 5

34. If f is a linear function and $f(-2) = 11$, $f(5) = -2$, and $f(x) = 4.3$, what is the value of x?

(A) –3.1
(B) –1.9
(C) 1.6
(D) 2.9
(E) 3.2

35. A taxicab company wanted to determine the fuel cost of its fleet. A sample of 30 vehicles was selected, and the fuel cost for the last month was tabulated for each vehicle. Later it was discovered that the highest amount was mistakenly recorded with an extra zero, so it was 10 times the actual amount. When the correction was made, this was still the highest amount. Which of the following must have remained the same after the correction was made?

(A) mean
(B) median
(C) mode
(D) range
(E) standard deviation

36. The range of the function $y = x^{-2/3}$ is

(A) $y < 0$
(B) $y > 0$
(C) $y \geq 0$
(D) $y \leq 0$
(E) all real numbers

37. The formula $A = Pe^{0.04t}$ gives the amount A that a savings account will be worth if an initial investment P is compounded continuously at an annual rate of 4 percent for t years. Under these conditions, how many years will it take an initial investment of $10,000 to be worth approximately $25,000?

(A) 1.9
(B) 2.5
(C) 9.9
(D) 22.9
(E) 25.2

38. A coin is tossed three times. Given that at least one head appears, what is the probability that exactly two heads will appear?

(A) $\dfrac{3}{8}$

(B) $\dfrac{3}{7}$

(C) $\dfrac{5}{8}$

(D) $\dfrac{3}{4}$

(E) $\dfrac{7}{8}$

39. A unit vector parallel to vector $\vec{V} = (2,-3,6)$ is vector

(A) $(-2,3,-6)$
(B) $(6,-3,2)$
(C) $(-0.29,0.43,-0.86)$
(D) $(0.29,0.43,-0.86)$
(E) $(-0.36,-0.54,1.08)$

40. What is the equation of the horizontal asymptote of the function $f(x) = \dfrac{(2x-1)(x+3)}{(x+3)^2}$?

(A) $y = -9$
(B) $y = -3$
(C) $y = 0$
(D) $y = \dfrac{1}{2}$
(E) $y = 2$

41. The points in the rectangular coordinate plane are transformed in such a way that each point $A(x,y)$ is moved to a point $A'(kx,ky)$. If the distance between a point A and the origin is d, then the distance between the origin and the point A' is

(A) $\dfrac{k}{d}$

(B) $\dfrac{d}{k}$

(C) d

(D) kd

(E) d^2

42. A committee of 5 people is to be selected from 6 men and 9 women. If the selection is made randomly, what is the probability that the committee consists of 3 men and 2 women?

(A) $\dfrac{1}{9}$

(B) $\dfrac{240}{1,001}$

(C) $\dfrac{1}{3}$

(D) $\dfrac{1,260}{3,003}$

(E) $\dfrac{13}{18}$

43. Three consecutive terms, in order, of an arithmetic sequence are $x + \sqrt{2}$, $2x + \sqrt{3}$, and $5x - \sqrt{5}$. Then x equals

(A) 2.14
(B) 2.45
(C) 2.46
(D) 3.24
(E) 3.56

44. The graph of $xy - 4x - 2y - 4 = 0$ can be expressed as a set of parametric equations. If $y = \dfrac{4t}{t-3}$ and $x = f(t)$, then $f(t) =$

(A) $t + 1$

(B) $t - 1$

(C) $3t - 3$

(D) $\dfrac{t-3}{4t}$

(E) $\dfrac{t-3}{2}$

45. If $f(x) = ax^2 + bx + c$, how must a and b be related so that the graph of $f(x - 3)$ will be symmetric about the y-axis?

(A) $a = b$
(B) $b = 0$, a is any real number
(C) $b = 3a$
(D) $b = 6a$
(E) $a = \dfrac{1}{9}b$

PRACTICE TEST 2

46. The graph of $y = \log_5 x$ and $y = \ln 0.5x$ intersect at a point where x equals

(A) 6.24
(B) 5.44
(C) 1.69
(D) 1.14
(E) 1.05

47. What is the value of x if $\pi \leq x \leq \dfrac{3\pi}{2}$ and $\sin x = 5 \cos x$?

(A) 3.399
(B) 3.625
(C) 4.515
(D) 4.623
(E) 4.663

48. The area of the region enclosed by the graph of the polar curve $r = \dfrac{1}{\sin\theta + \cos\theta}$ and the x- and y-axes is

(A) 0.48
(B) 0.50
(C) 0.52
(D) 0.98
(E) 1.00

49. A rectangular box has dimensions of length = 6, width = 4, and height = 5. The measure of the angle formed by a diagonal of the box with the base of the box is

(A) 27°
(B) 35°
(C) 40°
(D) 44°
(E) 55°

50. If (x, y) represents a point on the graph of $y = 2x + 1$, which of the following could be a portion of the graph of the set of points (x, y^2)?

(A)

(B)

(C)

(D)

(E)

STOP

If there is still time remaining, you may review your answers.

1. E	14. D	27. E	40. E
2. B	15. A	28. B	41. D
3. E	16. C	29. D	42. B
4. D	17. B	30. D	43. A
5. A	18. E	31. C	44. B
6. D	19. C	32. D	45. D
7. B	20. E	33. B	46. A
8. E	21. C	34. C	47. C
9. A	22. C	35. B	48. B
10. C	23. A	36. B	49. B
11. D	24. B	37. D	50. C
12. C	25. A	38. B	
13. A	26. E	39. C	

ANSWERS EXPLAINED

The following explanations are keyed to the review portions of this book. The number in brackets after each explanation indicates the appropriate chapter in the Review of Major Topics (Part 2).

An asterisk appears next to those solutions for which a graphing calculator is necessary or helpful.

* 1. **(E)** Graph the function f in the standard window and observe the vertical asymptote at $x = -2$.

 An alternative solution is to factor the denominator of f as $(x + 2)(x - 2)$; cancel the factor $x - 2$ in the numerator and in the denominator so that $f(x) = \dfrac{1}{x+2}$; and recall that a function has a vertical asymptote when the denominator is zero and the numerator isn't. There is a hole in the graph of f at $x = 2$. [9]

* 2. **(B)** Use the distance formula for three-dimensional space:
$$d = \sqrt{(x_2 - x_1)^2 + (y_2 - y_1)^2 + (z_2 - z_1)^2} = \sqrt{(2+3)^2 + (7-4)^2 + (-4-1)^2} \approx 7.68 \ [15]$$

3. **(E)** $\text{Log}(a^2 - b^2) = \log(a + b)(a - b) = \log(a + b) + \log(a - b)$. [8]

4. **(D)** In order that $\sqrt{x^2 - 9}$ be a real number, $x^2 - 9 \geq 0$. Solve this inequality to find that $x \leq -3$ or $x \geq 3$. [1]

5. **(A)** The slope of the first line is $-\dfrac{1}{2}$, and the slope of the second is $-\dfrac{a}{3}$. To be perpendicular, $-\dfrac{1}{2} = \dfrac{3}{a}$, or $a = -6$. [2]

* 6. **(D)** Plot the graph of $6 \sin x \cos x$ using ZTrig and observe that the max of this function is 3. (None of the other answer choices are close. If one were, you could use CALC/max to find the maximum value of the function.)

 An alternative solution is to recall that $\sin(2x) = 2 \sin x \cos x$, so that $6 \sin x \cos x = 3 \sin 2x$. Since the amplitude of $\sin 2x$ is 1, the amplitude of $3 \sin 2x$ is 3. [7]

* 7. **(B)** Put your calculator in radian mode. $f(2,3) = 2 \cdot \cos 3 \approx 2(-0.98999) \approx -1.98$. [14]

8. **(E)** Since 5 and -1 are zeros, $x - 5$ and $x + 1$ are factors of $P(x)$, so their product $x^2 - 4x - 5$ is too. [3]

* 9. **(A)** Enter the expression into your graphing calculator. You could also recall that the powers of i cycle in groups of four: $i, -1, -i, 1$. Therefore, the sum of any four consecutive powers of i is 0. [17]

10. **(C)** The length of an arc s is $s = r\theta$, and the area A of the sector formed by that arc is $A = \dfrac{1}{2}r^2\theta = \dfrac{1}{2}r(r\theta)$. Therefore, $8\pi = \dfrac{1}{2}r(4\pi)$, so $r = 4$. [7]

11. **(D)** The third and seventh terms are 4 terms apart, and the difference between them is 8. Therefore, the common difference is $8 \div 4 = 2$. Since the first term is two terms prior to the third term, its value is $15 - 2(2) = 11$. [19]

12. **(C)** Surface area $= 4\pi r^2$. Volume $= \dfrac{4}{3}\pi r^3$. Solve $4\pi r^2 = \dfrac{4}{3}\pi r^3$ to get $r = 3$. [15]

13. **(A)** The largest power of x "dominates" the other terms. This means that the value of the given expression gets closer to the value of $\dfrac{x^4}{x^5} = \dfrac{1}{x}$, which gets closer to 0 as x gets larger. [9]

14. **(D)** $s = r\,\theta$. $12 = r$. [7]

* **15.** **(A)** The domain consists of all numbers that make $3x^3 - 7 \geq 0$. Therefore,

$x^3 \geq \dfrac{7}{3}$ and $x \geq 1.33$. [1]

16. **(C)** Cofunctions of complementary angles are equal. Since x and y are complementary, tan and cot are cofunctions.

An alternative solution is to choose any values of x and y such that their sum is 90°. (For example, $x = 40°$ and $y = 50°$.) Test the answer choices with your calculator in degree mode to see that only Choice C is true. [7]

* **17.** **(B)** Plot the graph of $y = x^3 + 5x + 1$ in the standard window and zoom in a couple of times to see that it crosses the x-axis only once. To make sure you are not missing anything, you should also plot the equation in a $[-1,1]$ by $[-1,1]$ window and a $[-100,100]$, by $[-100,100]$ window. [4]

* **18.** **(E)** Complete the square in x and y: $(x^2 + 2x + 1) + (y^2 - 4y + 4) + z^2 = 10 + 1 + 4$, so $(x + 1)^2 + (y - 2)^2 + z^2 = 15$. Therefore, $r = \sqrt{15} \approx 3.87$. [15]

19. **(C)** If you substitute 0 for x you get c, so c is the y-intercept of the graph. The graph and window indicate that this value is about 80. [4]

* **20.** **(E)** Plot the graph of $y = x(x - 3)(x + 2)$ in the standard window and observe that the graph is above the x-axis when $-2 < x < 0$ or when $x > 3$.

An alternative solution is to find the zeros of the function $x(x - 3)(x + 2)$ as $x = 0, 3, -2$ and test points in the intervals established by these zeros. Points between -2 and 0 and greater than 3 satisfy the inequality. [5]

* **21.** **(C)** Use the two-dimensional version of the formula on page 345 to calculate the distance between the point $(0,0)$ and the line $3x - 4y = 10$ as 2. Therefore, the radius of the circle is 2, and its equation is $x^2 + y^2 = 4$. [13]

22. **(C)**

$$\frac{f(x+t)-f(x)}{t} = \frac{3-2(x+t)+(x+t)^2-(3-2x+x^2)}{t}$$

$$= \frac{3-2x-2t+x^2+2xt+t^2-3+2x-x^2}{t}$$

$$= 2x-2+t \quad [1]$$

> **CAUTION:** For calculus students only: This difference quotient looks like the definition of the derivative. However, no limit is taken, so don't jump at $f'(x)$, which is Choice D.

* 23. **(A)** Enter x^3 into Y_1 and x^2+1 into Y_2. Then enter $Y_1 Y_2$ into Y_3; $Y_1(Y_2)$ into Y_4; and $Y_2(Y_1)$ into Y_5. De-select Y_1 and Y_2. Inspection of Y_3, Y_4, and Y_5 shows that only Y_3 is symmetric about the origin.

An alternative solution is to define each of the three functions as $h(x)$ and check each against the definition of an odd function, $h(-x) = -h(x)$:

$$h(-x) = (-x)^3((-x)^2+1) = -x^5-x^3 = -(x^5+x^3)$$
$$= -h(x)$$
$$h(-x) = f((-x)^2+1) = (x^2+1)^3 \neq -h(x)$$
$$h(-x) = g((-x)^3) = ((-x)^3)^2+1 = x^6+1 \neq -h(x) \quad [1]$$

* 24. **(B)** Since one particular man must be on the committee, the problem becomes: "Form a committee of 3 from 8 men." Calculate $8_nC_r3 = 56$. [16]

25. **(A)** Each rectangle has width 1. The heights of these rectangles are $2^1 = 2$, $2^2 = 4$, and $2^3 = 8$. The sum of these areas is 14. [8]

* 26. **(E)** Mean $= \dfrac{1+2+3+1+2+5+x}{7} = \dfrac{14+x}{7} = 3.\overline{27}$. Therefore, $x = 8.9$. [21]

* 27. **(E)** Law of Sines:

$$\frac{\frac{1}{3}}{\sqrt{5}} = \frac{\frac{3}{5}}{KL}$$

$$\frac{1}{3}KL = \frac{3\sqrt{5}}{5}$$

$$KL = \frac{9\sqrt{5}}{5} \approx 4.0 \quad [6]$$

28. **(B)** Multiplication of matrices is possible only when the number of columns in the matrix on the left equals the number of rows in the matrix on the right. In that case, the product is a matrix with the number of rows in the left-hand matrix and the number of columns of the right-hand matrix. [18]

29. **(D)** Add $-1 + 5x$ to both sides to get $x^2 + 5x + 4 > 0$. Factor the left side to get $(x + 4)(x + 1) > 0$, and observe that the zeros of the corresponding equation are -4 and -1. Using $-5, -2$, and 0 as test points, find that $(x + 4)(x + 1) > 0$ when $x < -4$ or $x > -1$. [5]

30. **(D)** $(g \circ f)(x) = g(f(x)) = \sqrt{x - 7}$. Therefore, $x \geq 7$. [1]

* 31. **(C)** Law of Cosines: $c^2 = 16 + 1 - 8 \cdot \dfrac{\sqrt{3}}{2} = 17 - 4\sqrt{3} \approx 10.07$. Therefore, $c \approx 3.2$. [6]

* 32. **(D)** Use the standard window to graph $y < -\dfrac{3}{4}x$, by moving the cursor all the way left (past Y =) and keying Enter until a "lower triangle" is observed. The shaded portion of the graph will lie in all but quadrant I.

    An alternative solution is to graph the related equation $y = -\dfrac{3}{4}x$ and test points to determine which side of the line contains solutions to the inequality. This will indicate the quadrant that the graph does not enter. [2]

33. **(B)** Graph the first term of f to see that $\dfrac{5}{x + 1}$ can be any value except 0. Therefore, $f(x)$ cannot be -4. [1]

* 34. **(C)** Since these points are on a line, all slopes must be equal.

    Slope $= \dfrac{-2 - 11}{5 - (-2)} = \dfrac{-13}{7} = \dfrac{4.3 - 11}{x - (-2)}$. Cross-multiplying the far right equation gives $-13(x + 2) = 7(-6.7)$, and solving for x yields $x \approx 1.6$. [2]

35. **(B)** Since the median is the middle value, it does not change if the number (quantity) of values above and below it remain the same. [21]

* 36. **(B)** Plot the graph of $y = x^{-2/3}$ in the standard window and observe that the entire graph lies above the x-axis.

    An alternative solution uses the fact that $x^{-2/3} = \dfrac{1}{x^{2/3}}$, and $x^{2/3} = (x^{1/3})^2$, so that for all values of x (except zero) y is positive. Therefore, the range of $y = x^{-2/3}$ is $y > 0$. [8]

* 37. **(D)** Substitute the given values into the formula to get $25{,}000 = 10{,}000e^{0.04t}$. To solve for t, first divide both sides by 10,000. Then take the ln of both sides to get $\ln 2.5 = 0.04t$. Therefore $t = \dfrac{\ln 2.5}{0.04} = 22.9$. [8]

38. **(B)** There are 8 elements in the sample space of a coin being flipped 3 times. Of these elements, 7 contain at least 1 head and 3 (HHT, HTH, THH) contain 2 heads. Probability $= \dfrac{3}{7}$. [22]

* 39. **(C)** To find a unit vector parallel to $(2, -3, 6)$, divide each component by $\sqrt{2^2 + (-3)^2 + 6^2} = \pm 7$. There are two unit vectors, pointing in opposite directions, that meet this requirement: $(0.29, -0.43, 0.86)$ and $(-0.29, 0.43, -0.86)$. Only the second one is an answer choice. [20]

40.   **(E)** To find the horizontal asymptote, you need to see what happens when x gets much larger ($\to \infty$) or much smaller ($\to -\infty$). With the numerator and the denominator expanded, $f(x) = \dfrac{2x^2 + 5x - 3}{x^2 + 6x + 9}$. Divide both by x^2 to see that y approaches 2 as x gets much bigger or smaller. [9]

41.   **(D)** The distance between the origin and A is $d = \sqrt{x^2 + y^2}$. The distance between the

origin and A' is $\sqrt{(kx)^2 + (ky)^2} = \sqrt{k^2(x^2 + y^2)} = k\sqrt{x^2 + y^2} = kd$. [12]

* 42.   **(B)** There are $\binom{15}{5}$ ways of selecting a committee of 5 out of 15 people (men and

women). There are $\binom{6}{3}$ ways of selecting 3 men out of 6, and for each of these, there

are $\binom{9}{2}$ ways of selecting 2 women out of 9. Therefore, there are $\binom{6}{3}\binom{9}{2}$ ways

of selecting 3 men and 2 women. The probability of selecting 3 men and 2 women is

$\dfrac{\binom{6}{3}\binom{9}{2}}{\binom{15}{5}} = \dfrac{240}{1{,}001}$. Using the calculator command $_nC_r = \binom{n}{r}$, enter this expression

and change it to a fraction. [22]

* 43.   **(A)** $2x + \sqrt{3} = (x + \sqrt{2}) + d$, and $5x - \sqrt{5} = (2x + \sqrt{3}) + d$. Eliminate d, and $x + \sqrt{3} - \sqrt{2} = 3x - \sqrt{5} - \sqrt{3}$. Thus, $x = \dfrac{2\sqrt{3} - \sqrt{2} + \sqrt{5}}{2} \approx 2.14$. [19]

44.   **(B)** Substituting for y gives $x\left(\dfrac{4t}{t-3}\right) - 4x - 2\left(\dfrac{4t}{t-3}\right) - 4 = 0$. Multiplying both sides by $t - 3$ yields $12x - 12t + 12 = 0$. Then $x = t - 1$. [10]

45.   **(D)** The graph of the function is a parabola, and the equation of its axis of symmetry is $x = -\dfrac{b}{2a}$. Since the graph of $y = f(x - 3)$ is the translation of $y = f(x)$ 3 units to the right, $f(x - 3)$ will be symmetric about the y-axis when $f(x)$ is symmetric about -3. Therefore, $-\dfrac{b}{2a} = -3$, or $b = 6a$. [3]

* 46. **(A)** Use the change-of-base formula and graph $y = \dfrac{\log x}{\log 5}$ as Y_1. Graph $y = \ln(0.5x)$ as Y_2. The answer choices suggest a $[0,10]$ by $[0,2]$ window. Use the CALC/intersect to find the correct answer, Choice A.

An alternative solution converts the equations to exponential form: $x = 5^y$ and $\dfrac{1}{2}x = e^y$, or $5^y = 2e^y$. Taking the natural log of both sides gives $y \ln 5 = \ln 2 + y$, and solving for y yields $y = \dfrac{\ln 2}{\ln 5 - 1} \approx 1.14$. Substitute back to find x: $5^{1.14} \approx 6.24$. [8]

* 47. **(C)** With your calculator in radian mode, plot the graphs of $Y_1 = \sin x$ and $Y_2 = 5\cos x$ in a $\left[\pi, \dfrac{3\pi}{2}\right]$ by $[-5,5]$ window. The solution is the x-coordinate of the point of intersection.

An alternative solution is to divide both sides of the equation by $\cos x$ to get $\tan x = 5$. Use your calculator to get $\tan^{-1} 5 = 1.373$. This is the first-quadrant solution, so add π. [7]

* 48. **(B)** With your calculator in polar mode, graph $r = \dfrac{1}{\sin\theta + \cos\theta}$ in a $[-3,3]$ by $[-3,3]$ window. Let θ run from 0 to 2π in increments of 0.1. The graph of the function and the two axes form an isosceles right triangle of side length 1, so its area is 0.5.

An alternative solution is to multiply both sides of the equation by $\sin\theta + \cos\theta$ to get $r\sin\theta + r\cos\theta = 1$. Since $x = r\cos\theta$ and $y = r\sin\theta$, the rectangular form of the equation is $x + y = 1$. This line makes an isosceles right triangle of unit leg lengths with the axes, the area of which is $\dfrac{1}{2}(1)(1) = 0.5$. [14]

* 49. **(B)** The diagonal of the base is $\sqrt{6^2 + 4^2} = \sqrt{52}$. The diagonal of the box is the hypotenuse of a right triangle with one leg $\sqrt{52}$ and the other leg 5. Let θ be the angle formed by the diagonal of the base and the diagonal of the box. Since $\tan\theta = \dfrac{5}{\sqrt{52}} \approx 0.69337$, then $\theta = \tan^{-1} 0.69337 \approx 35°$. [6]

* 50. **(C)** Plot $y = (2x + 1)^2$ in a window with a $[-2,2]$ by $[-1,5]$ window and observe that Choice C is the only possible choice.

An alternative solution is to note that the graph of $y = (2x+1)^2 = 4\left(x + \dfrac{1}{2}\right)^2$ is a parabola with a vertex at $\left(-\dfrac{1}{2}, 0\right)$ that opens up. The only answer choice with these properties is Choice C. [2]

Self-Evaluation Chart for Practice Test 2

Subject Area	Questions and Review Section							Right	Number Wrong	Omitted
Algebra and Functions (25 questions)	1 9	3 8	4 1	5 2	8 3	13 9		_____	_____	_____
	15 1	17 4	19 4	20 5	22 1	23 1		_____	_____	_____
	25 8	29 5	30 1	32 2	33 1	34 2	36 8	_____	_____	_____
	37 8	40 9	44 10	45 3	46 8	50 2		_____	_____	_____
Trigonometry (8 questions)	6 7	10 7	14 7	16 7	27 6	31 6		_____	_____	_____
	47 7	49 6						_____	_____	_____
Coordinate and Three-Dimensional Geometry (7 questions)	2 15	7 14	12 15	18 15	21 13	41 12	48 14	_____	_____	_____
Number and Operations (6 questions)	9 17	11 19	24 16	28 18	39 20			_____	_____	_____
	43 19							_____	_____	_____
Data Analysis, Statistics, and Probability (4 questions)	26 21	35 21	38 22	42 22				_____	_____	_____
TOTALS								_____	_____	_____

Evaluate Your Performance on Practice Test 2

Rating	Number Right
Excellent	41–50
Very good	33–40
Above average	25–32
Average	15–24
Below average	Below 15

CALCULATING YOUR SCORE

Raw score R = number right $- \dfrac{1}{4}$(number wrong), rounded = _____

Approximate scaled score $S = 800 - 10(44 - R) = $ _____

If $R \geq 44$, $S = 800$.

ANSWER SHEET
Practice Test 3

1. Ⓐ Ⓑ Ⓒ Ⓓ Ⓔ
2. Ⓐ Ⓑ Ⓒ Ⓓ Ⓔ
3. Ⓐ Ⓑ Ⓒ Ⓓ Ⓔ
4. Ⓐ Ⓑ Ⓒ Ⓓ Ⓔ
5. Ⓐ Ⓑ Ⓒ Ⓓ Ⓔ
6. Ⓐ Ⓑ Ⓒ Ⓓ Ⓔ
7. Ⓐ Ⓑ Ⓒ Ⓓ Ⓔ
8. Ⓐ Ⓑ Ⓒ Ⓓ Ⓔ
9. Ⓐ Ⓑ Ⓒ Ⓓ Ⓔ
10. Ⓐ Ⓑ Ⓒ Ⓓ Ⓔ
11. Ⓐ Ⓑ Ⓒ Ⓓ Ⓔ
12. Ⓐ Ⓑ Ⓒ Ⓓ Ⓔ
13. Ⓐ Ⓑ Ⓒ Ⓓ Ⓔ

14. Ⓐ Ⓑ Ⓒ Ⓓ Ⓔ
15. Ⓐ Ⓑ Ⓒ Ⓓ Ⓔ
16. Ⓐ Ⓑ Ⓒ Ⓓ Ⓔ
17. Ⓐ Ⓑ Ⓒ Ⓓ Ⓔ
18. Ⓐ Ⓑ Ⓒ Ⓓ Ⓔ
19. Ⓐ Ⓑ Ⓒ Ⓓ Ⓔ
20. Ⓐ Ⓑ Ⓒ Ⓓ Ⓔ
21. Ⓐ Ⓑ Ⓒ Ⓓ Ⓔ
22. Ⓐ Ⓑ Ⓒ Ⓓ Ⓔ
23. Ⓐ Ⓑ Ⓒ Ⓓ Ⓔ
24. Ⓐ Ⓑ Ⓒ Ⓓ Ⓔ
25. Ⓐ Ⓑ Ⓒ Ⓓ Ⓔ
26. Ⓐ Ⓑ Ⓒ Ⓓ Ⓔ

27. Ⓐ Ⓑ Ⓒ Ⓓ Ⓔ
28. Ⓐ Ⓑ Ⓒ Ⓓ Ⓔ
29. Ⓐ Ⓑ Ⓒ Ⓓ Ⓔ
30. Ⓐ Ⓑ Ⓒ Ⓓ Ⓔ
31. Ⓐ Ⓑ Ⓒ Ⓓ Ⓔ
32. Ⓐ Ⓑ Ⓒ Ⓓ Ⓔ
33. Ⓐ Ⓑ Ⓒ Ⓓ Ⓔ
34. Ⓐ Ⓑ Ⓒ Ⓓ Ⓔ
35. Ⓐ Ⓑ Ⓒ Ⓓ Ⓔ
36. Ⓐ Ⓑ Ⓒ Ⓓ Ⓔ
37. Ⓐ Ⓑ Ⓒ Ⓓ Ⓔ
38. Ⓐ Ⓑ Ⓒ Ⓓ Ⓔ
39. Ⓐ Ⓑ Ⓒ Ⓓ Ⓔ

40. Ⓐ Ⓑ Ⓒ Ⓓ Ⓔ
41. Ⓐ Ⓑ Ⓒ Ⓓ Ⓔ
42. Ⓐ Ⓑ Ⓒ Ⓓ Ⓔ
43. Ⓐ Ⓑ Ⓒ Ⓓ Ⓔ
44. Ⓐ Ⓑ Ⓒ Ⓓ Ⓔ
45. Ⓐ Ⓑ Ⓒ Ⓓ Ⓔ
46. Ⓐ Ⓑ Ⓒ Ⓓ Ⓔ
47. Ⓐ Ⓑ Ⓒ Ⓓ Ⓔ
48. Ⓐ Ⓑ Ⓒ Ⓓ Ⓔ
49. Ⓐ Ⓑ Ⓒ Ⓓ Ⓔ
50. Ⓐ Ⓑ Ⓒ Ⓓ Ⓔ

Practice Test 3

T ear out the preceding answer sheet. Decide which is the best choice by rounding your answer when appropriate. Blacken the corresponding space on the answer sheet. When finished, check your answers with those at the end of the test. For questions that you got wrong, note the chapters containing the material that you must review. Also, if you do not fully understand how you arrived at some of the correct answers, you should review the appropriate chapters. Finally, fill out the Self-Evaluation Chart on page 275 in order to pinpoint the topics that gave you the most difficulty.

50 questions: 1 hour

Directions: Decide which answer choice is best. If the exact numerical value is not one of the answer choices, select the closest approximation. Fill in the circle on the answer sheet that corresponds to your choice.

Notes:
(1) You will need to use a scientific or graphing calculator to answer some of the questions.
(2) You will have to decide whether to put your calculator in degree or radian mode for some problems.
(3) All figures that accompany problems are plane figures unless otherwise stated. Figures are drawn as accurately as possible to provide useful information for solving the problem, except when it is stated in a particular problem that the figure is not drawn to scale.
(4) Unless otherwise indicated, the domain of a function is the set of all real numbers for which the functional value is also a real number.

Reference Information. The following formulas are provided for your information.

Volume of a right circular cone with radius r and height h: $V = \frac{1}{3}\pi r^2 h$

Volume of a sphere of radius r: $V = \frac{4}{3}\pi r^3$

Surface area of a sphere of radius r: $S = 4\pi r^2$

Volume of a pyramid of base area B and height h: $V = \frac{1}{3}Bh$

1. The slope of a line that is perpendicular to $3x + 2y = 7$ is

(A) -2

(B) $-\dfrac{3}{2}$

(C) $\dfrac{2}{3}$

(D) $\dfrac{3}{2}$

(E) 2

2. What is the remainder when $3x^4 - 2x^3 - 20x^2 - 12$ is divided by $x + 2$?

(A) -60
(B) -36
(C) -28
(D) -6
(E) -4

3. If $f(x) = 3x$ and $g(x) = \cos x$, then $(g \circ f)\left(\dfrac{\pi}{3}\right) =$

(A) -1
(B) 1

(C) $\dfrac{3}{2}$

(D) $\dfrac{3\pi}{2}$

(E) 3

4. If $f(x) = 2 \ln x + 3$ and $g(x) = e^x$, then $f(g(3)) =$

(A) 9
(B) 11
(C) 43.17
(D) 47.13
(E) 180.77

5. The domain of $f(x) = \log(\sin x)$ contains which of the following intervals?

(A) $0 \leq x \leq \pi$

(B) $-\dfrac{\pi}{2} \leq x \leq \dfrac{\pi}{2}$

(C) $0 < x < \pi$

(D) $-\dfrac{\pi}{2} < x < \dfrac{\pi}{2}$

(E) $\dfrac{\pi}{2} < x < \dfrac{3\pi}{2}$

6. Which of the following is the ratio of the surface area of the sphere with radius r to its volume?

(A) $\dfrac{4}{r}$

(B) $\dfrac{3}{r}$

(C) $\dfrac{r}{4}$

(D) $\dfrac{r}{\pi}$

(E) $\dfrac{4}{\pi}$

7. If the two solutions of $x^2 - 9x + c = 0$ are complex conjugates, which of the following describes all possible values of c?

(A) $c = 0$

(B) $c \neq 0$

(C) $c < 9$

(D) $c > \dfrac{81}{4}$

(E) $c > 81$

8. If $\tan x = 3$, the numerical value of $\sqrt{\csc x}$ is

(A) 0.32
(B) 0.97
(C) 1.03
(D) 1.78
(E) 3.16

9. In the figure above, the graph of $y = f(x)$ has two transformations performed on it. First it is rotated 180° about the origin, and then it is reflected about the x-axis. Which of the following is the equation of the resulting curve?

(A) $y = -f(x)$
(B) $y = f(x + 2)$
(C) $x = f(y)$
(D) $y = f(x)$
(E) none of the above

10. If $f(x) = \dfrac{3x^3 - 7x^2 + 2}{4x^2 - 3x - 1}$, what does $f(x)$ approach as x gets infinitely larger?

    (A) 0

    (B) $\dfrac{3}{4}$

    (C) 1

    (D) 3

    (E) ∞

11. The set of points (x,y,z) such that $x = 5$ is

    (A) a point
    (B) a line
    (C) a plane
    (D) a circle
    (E) a cube

12. The vertical distance between the minimum and maximum values of the function $y = \left| -\sqrt{2} \sin \sqrt{3x} \right|$ is

    (A) 1.414
    (B) 1.732
    (C) 2.094
    (D) 2.828
    (E) 3.464

13. If the domain of $f(x) = -|x| + 2$ is $-1 \leq x \leq 3$, $f(x)$ has a minimum value when x equals

    (A) -1
    (B) 0
    (C) 1
    (D) 3
    (E) There is no minimum value.

14. What is the range of the function $f(x) = x^2 - 14x + 43$?

    (A) $x \leq 7$
    (B) $x \geq 0$
    (C) $y \leq -6$
    (D) $y \geq -6$
    (E) all real numbers

15. A positive rational root of the equation
$4x^3 - x^2 + 16x - 4 = 0$ is

(A) $\dfrac{1}{4}$

(B) $\dfrac{1}{2}$

(C) $\dfrac{3}{4}$

(D) 1

(E) 2

16. The norm of vector $\vec{V} = 3\vec{i} - \sqrt{2}\,\vec{j}$ is

(A) 4.24
(B) 3.61
(C) 3.32
(D) 2.45
(E) 1.59

17. If five coins are flipped and all the different ways they could fall are listed, how many elements of this list will contain more than 3 heads?

(A) 5
(B) 6
(C) 10
(D) 16
(E) 32

18. The seventh term of an arithmetic sequence is 5 and the twelfth term is –15. The first term of this sequence is

(A) 28
(B) 29
(C) 30
(D) 31
(E) 32

19. The graph of the curve represented by $\begin{cases} x = \sec\theta \\ y = \cos\theta \end{cases}$ is

(A) a line
(B) a hyperbola
(C) an ellipse
(D) a line segment
(E) a portion of a hyperbola

20. Point (3,2) lies on the graph of the inverse of
$f(x) = 2x^3 + x + A$. The value of A is

(A) –54
(B) –15
(C) 15
(D) 18
(E) 54

21. If $f(x) = ax^2 + bx + c$ and $f(1) = 3$ and $f(-1) = 3$, then $a + c$ equals

(A) –3
(B) 0
(C) 2
(D) 3
(E) 6

USE THIS SPACE FOR SCRATCH WORK

22. In $\triangle ABC$, $\angle B = 42°$, $\angle C = 30°$, and $AB = 100$. The length of $\overline{BC}$ is

(A) 47.6
(B) 66.9
(C) 133.8
(D) 190.2
(E) 193.7

23. If $4\sin x + 3 = 0$ on $0 \leq x < 2\pi$, then $x =$

(A) –0.848
(B) 0.848
(C) 5.435
(D) 0.848 or 5.435
(E) 3.990 or 5.435

24. What is the sum of the infinite geometric series

$$6 + 4 + \frac{8}{3} + \frac{16}{9} + \ldots?$$

(A) 18
(B) 36
(C) 45
(D) 60
(E) There is no sum.

25. In $a + bi$ form, the reciprocal of $2 + 6i$ is

(A) $\frac{1}{2} + \frac{1}{6}i$

(B) $-\frac{1}{16} + \frac{3}{16}i$

(C) $\frac{1}{16} + \frac{3}{16}i$

(D) $\frac{1}{20} - \frac{3}{20}i$

(E) $\frac{1}{20} + \frac{3}{20}i$

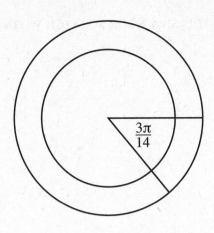

26. A central angle of two concentric circles is $\dfrac{3\pi}{14}$.

The area of the large sector is twice the area of the small sector. What is the ratio of the lengths of the radii of the two circles?

(A) 0.25 : 1
(B) 0.50 : 1
(C) 0.67 : 1
(D) 0.71 : 1
(E) 1 : 1

27. If the region bounded by the lines $y = -\dfrac{4}{3}x + 4$,

$x = 0$, and $y = 0$ is rotated about the y-axis, the volume of the figure formed is

(A) 18.8
(B) 37.7
(C) 56.5
(D) 84.8
(E) 113.1

28. If there are known to be 4 broken transistors in a box of 12, and 3 transistors are drawn at random, what is the probability that none of the 3 is broken?

(A) 0.250
(B) 0.255
(C) 0.375
(D) 0.556
(E) 0.750

29. What is the domain of $f(x) = \sqrt[3]{15 - x^2}$?

(A) $x > 0$
(B) $x > 2.47$
(C) $-2.47 < x < 2.47$
(D) $-3.87 < x < 3.87$
(E) all real numbers

30. Which of the following is a horizontal asymptote to

the function $f(x) = \dfrac{3x^4 - 7x^3 + 2x^2 + 1}{2x^4 - 4}$?

(A) $y = -3.5$
(B) $y = 0$
(C) $y = 0.25$
(D) $y = 0.75$
(E) $y = 1.5$

31. When a certain radioactive element decays, the amount at any time t can be calculated using the

function $E(t) = ae^{\frac{-t}{500}}$, where a is the original amount and t is the elapsed time in years. How many years would it take for an initial amount of 250 milligrams of this element to decay to 100 milligrams?

(A) 125 years
(B) 200 years
(C) 458 years
(D) 496 years
(E) 552 years

32. If n is an integer, what is the remainder when $3x^{(2n+3)} - 4x^{(2n+2)} + 5x^{(2n+1)} - 8$ is divided by $x + 1$?

(A) -20
(B) -10
(C) -4
(D) 0
(E) The remainder cannot be determined.

33. Four men, A, B, C, and D, line up in a row. What is the probability that man A is at either end of the row?

(A) $\dfrac{1}{2}$

(B) $\dfrac{1}{3}$

(C) $\dfrac{1}{4}$

(D) $\dfrac{1}{6}$

(E) $\dfrac{1}{12}$

34. $\displaystyle\sum_{i=3}^{10} 5 =$

(A) 260
(B) 50
(C) 40
(D) 5
(E) none of the above

35. The graph of $y^4 - 3x^2 + 7 = 0$ is symmetric with respect to which of the following?

  I. the x-axis
 II. the y-axis
III. the origin

(A) only I
(B) only II
(C) only III
(D) only I and II
(E) I, II, and III

36. In a group of 30 students, 20 take French, 15 take Spanish, and 5 take neither language. How many students take both French and Spanish?

(A) 0
(B) 5
(C) 10
(D) 15
(E) 20

37. If $f(x) = x^2$, then $\dfrac{f(x+h) - f(x)}{h} =$

(A) 0
(B) h
(C) $2x$
(D) $2x + h$
(E) $\dfrac{x^2}{h}$

38. The plane whose equation is $5x + 6y + 10z = 30$ forms a pyramid in the first octant with the coordinate planes. Its volume is

(A) 15
(B) 21
(C) 30
(D) 36
(E) 45

39. What is the range of the function $f(x) = \dfrac{3}{x-5} - 1$?

   (A) All real numbers
   (B) All real numbers except 5
   (C) All real numbers except 0
   (D) All real numbers except –1
   (E) All real numbers greater than 5

40. You are given the set of data $1, 1, 2, 2, 2, 3, 3, x, y$, where x and y represent two different integers. If the mode is 2, which of the following statements must be true?

   (A) If $x = 1$ or 3, then y must $= 2$.
   (B) Both x and y must be > 3.
   (C) Either x or y must $= 2$.
   (D) It does not matter what values x and y have.
   (E) Either x or y must $= 3$, and the other must $= 1$.

41. If $f(x) = \sqrt{2x+3}$ and $g(x) = x^2$, for what value(s) of x does $f(g(x)) = g(f(x))$?

   (A) –0.55
   (B) 0.46
   (C) 5.45
   (D) –0.55 and 5.45
   (E) 0.46 and 6.46

42. If $3x - x^2 \geq 2$ and $y^2 + y \leq 2$, then

   (A) $-1 \leq xy \leq 2$
   (B) $-2 \leq xy \leq 2$
   (C) $-4 \leq xy \leq 4$
   (D) $-4 \leq xy \leq 2$
   (E) $xy = 1, 2,$ or 4 only

43. In $\triangle ABC$, if $\sin A = \dfrac{1}{3}$ and $\sin B = \dfrac{1}{4}$, $\sin C =$

   (A) 0.14
   (B) 0.54
   (C) 0.56
   (D) 3.15
   (E) 2.51

PRACTICE TEST 3

44. The solution set of $\dfrac{|x-1|}{x} > 2$ is

(A) $0 < x < \dfrac{1}{3}$

(B) $x < \dfrac{1}{3}$

(C) $x > \dfrac{1}{3}$

(D) $\dfrac{1}{3} < x < 1$

(E) $x > 0$

45. Suppose the graph of $f(x) = -x^3 + 2$ is translated 2 units right and 3 units down. If the result is the graph of $y = g(x)$, what is the value of $g(-1.2)$?

(A) -33.77
(B) -1.51
(C) -0.49
(D) 31.77
(E) 37.77

46. In the figure above, the bases, *ABC* and *DEF*, of the right prism are equilateral triangles. The altitude of the prism is *BE*. If a plane cuts the figure through points *A, C*, and *E*, two solids, *EABC*, and *EACFD*, are formed. What is the ratio of the volume of *EABC* to the volume of *EACFD*?

(A) $\dfrac{1}{4}$

(B) $\dfrac{1}{3}$

(C) $\dfrac{\sqrt{3}}{4}$

(D) $\dfrac{1}{2}$

(E) $\dfrac{\sqrt{3}}{3}$

47. The range of the piecewise function

$$f(x) = \begin{cases} 3(x-1)^2 - 2 & \text{if } x < 6 \\ -2x + 5 & \text{if } x \geq 6 \end{cases}$$

in interval notation is

(A) $(-\infty, \infty)$
(B) $(-\infty, -7]$
(C) $[-2, \infty)$
(D) $(-7, -2)$
(E) $(-\infty, -7] \cup [-2, \infty)$

48. The length of the major axis of the ellipse $3x^2 + 2y^2 - 6x + 8y - 1 = 0$ is

(A) $\sqrt{3}$

(B) $\sqrt{6}$

(C) $2\sqrt{3}$

(D) 4

(E) $2\sqrt{6}$

49. A recent survey reported that 60 percent of the students at a high school are girls and 65 percent of the girls at this high school play a sport. If a student at this high school was selected at random, what is the probability that the student is a girl who plays a sport?

(A) 0.10
(B) 0.21
(C) 0.32
(D) 0.39
(E) 0.42

50. If $x - 7$ divides $x^3 - 3k^3x^2 - 13x - 7$, then $k =$

(A) 1.19
(B) 1.34
(C) 1.72
(D) 4.63
(E) 5.04

STOP

If there is still time remaining, you may review your answers.

1. C	14. D	27. B	40. A
2. C	15. A	28. B	41. A
3. A	16. C	29. E	42. D
4. A	17. B	30. E	43. C
5. C	18. B	31. C	44. A
6. B	19. E	32. A	45. D
7. D	20. B	33. A	46. D
8. C	21. D	34. C	47. E
9. D	22. D	35. E	48. E
10. E	23. E	36. C	49. D
11. C	24. A	37. D	50. A
12. A	25. D	38. A	
13. D	26. D	39. D	

ANSWERS EXPLAINED

The following explanations are keyed to the review portions of this book. The number in brackets after each explanation indicates the appropriate chapter in the Review of Major Topics (Part 2).

An asterisk appears next to those solutions for which a graphing calculator is necessary or helpful.

1. **(C)** Transform the given equation into slope-intercept form: $y = -\frac{3}{2}x + \frac{7}{2}$ to see that the slope is $-\frac{3}{2}$. The slope of a perpendicular line is the negative reciprocal, or $\frac{2}{3}$. [2]

* 2. **(C)** By the Remainder Theorem, substituting -2 for x in $3x^4 - 2x^3 - 20x^2 - 12$ yields the remainder upon division by $x + 2$. [4]

3. **(A)** $(g \circ f)\left(\frac{\pi}{3}\right) = g\left(f\left(\frac{\pi}{3}\right)\right) = g(\pi) = \cos \pi = -1$. [1]

4. **(A)** Since $g(3) = e^3$, $f(g(3)) = 2 \ln e^3 + 3$. $\ln e^3 = 3$. So $f(g(3)) = 6 + 3 = 9$. [1]

5. **(C)** Since the domain of $\log_{10}$ is positive numbers, then the domain of f consists of values of x for which $\sin x$ is positive. This is only true for $0 < x < \pi$. [7]

6. **(B)** $\dfrac{\text{Surface area of sphere}}{\text{Volume of sphere}} = \dfrac{4\pi r^2}{\frac{4}{3}\pi r^3} = \dfrac{12}{4r} = \dfrac{3}{r}$. [15]

* 7. **(D)** Plot the graph of $y = x^2 - 9x$ in the standard window and observe that you must extend the window in the negative y direction to capture the vertex of the parabola. Since this vertex must lie above the x-axis for the solutions to be complex conjugates, c must be bigger than $|\text{minimum}| = 20.25 = \dfrac{81}{4}$. (The minimum is found using CALC/minimum.)

   An alternative solution is to use the fact that for the solutions to be complex conjugates, the discriminant $b^2 - 4ac = 81 - 4c < 0$, or $c > \dfrac{81}{4}$. [3]

* 8. **(C)** Since $\tan x = 3$, x could be in the first or third quadrants. Since, however, $\sqrt{\csc x}$ is only defined when $\csc x \geq 0$, we need only consider x in the first quadrant. Thus, we can enter $\sqrt{\left(\dfrac{1}{(\sin(\tan^{-1} 3))}\right)}$ to get the correct answer, Choice C. [7]

9. **(D)** The two transformations put the graph right back where it started. [12]

* 10. **(E)** Enter $(3x^3 - 7x^2 + 2)/(4x^2 - 3x - 1)$ into Y_1. Enter TBLSET and set TblStart $= 110$ and ΔTbl $= 10$. Then enter TABLE and scroll down to larger and larger x values until you are convinced that Y_1 grows without bound.

   An alternative solution is to divide the numerator and the denominator by x^3 and then let $x \to \infty$. The numerator approaches 3 while the denominator approaches 0, so the whole fraction grows without bound. [9]

11. **(C)** Since the y- and z-coordinates can have any values, the equation $x = 5$ is a plane where all points have an x-coordinate of 5. [15]

∗ 12. **(A)** Plot the graph of $y = \left| -\sqrt{2} \sin \sqrt{3}x \right|$ using Ztrig. The minimum value of the function is clearly zero, and you can use CALC/maximum to establish 1.414 as the maximum value.

This function inside the absolute value is sinusoidal with amplitude $\sqrt{2} \approx 1.414$. The absolute value eliminates the bottom portion of the sinusoid, so this is the vertical distance between the maximum and the minimum as well. [7]

∗ 13. **(D)** Plot the graph of $y = -|x| + 2$ in a $[-1,3]$ by $[-3,3]$ window. Examine the graph to see that its minimum value is achieved when $x = 3$.

An alternative solution is to realize that y is smallest when x is largest because of the negative absolute value. [11]

∗ 14. **(D)** Plot the graph of $y = x^2 - 14x + 43$, and find its minimum value of -6. All values of y greater than or equal to -6 are in the range of f. [1]

∗ 15. **(A)** Plot the graph of $y = 4x^3 - x^2 + 16x - 4$ in the standard window and zoom in once to get a clearer picture of the location of the zero. Use CALC/zero to determine that the zero is at $x = 0.25$.

An alternative solution is to use the Rational Roots Theorem to determine that the only possible rational roots are $\pm 1, \pm 2, \pm 4, \pm \frac{1}{2}, \frac{1}{4}$. Synthetic division with these values in turn will eventually yield the correct answer choice.

Another alternative solution is to observe that the left side of the equation can be factored: $4x^3 - x^2 + 16x - 4 = x^2(4x - 1) + 4(4x - 1) = (4x - 1)(x^2 + 1) = 0$. Since $x^2 + 1$ can never equal zero, the only solution is $x = \frac{1}{4}$. [4]

∗ 16. **(C)** $\left| \vec{v} \right| = \sqrt{3^2 + \left(-\sqrt{2}\right)^2} = \sqrt{9 + 2} = \sqrt{11} \approx 3.32$. [20]

∗ 17. **(B)** "More than 3" implies 4 or 5, and so the number of elements is $\binom{5}{4} + \binom{5}{5} = 6$. [16]

18. **(B)** There are 5 common differences, d, between the seventh and twelfth terms, so $5d = -15 - 5 = -20$, and $d = -4$. The seventh term is the first term plus $6d$, so $5 = t_1 - 24$, and $t_1 = 29$. [19]

∗ 19. **(E)** In parametric mode, plot the graph of $x = \sec t$ and $y = \cos t$ in the standard window to see that it looks like a portion of a hyperbola. You should verify that it is a portion of a hyperbola rather than the whole hyperbola by noting that $y = \cos t$ implies $-1 \leq y \leq 1$.

An alternative solution is to use the fact that secant and cosine are reciprocals so that elimination of the parameter t yields the equation $xy = 1$. This is the equation of a hyperbola and again, since $y = \cos t$ implies $-1 \leq y \leq 1$, the correct answer is Choice E. [10]

20. **(B)** If (3,2) lies on the inverse of f, (2,3) lies on f. Substituting in f gives $2 \cdot 2^3 + 2 + A = 3$. Therefore, $A = -15$. [1]

21. **(D)** Substitute 1 for x to get $a + b + c = 3$. Substitute -1 for x to get $a - b + c = 3$. Add these two equations to get $a + c = 3$. [1]

* **22.** **(D)** $\angle A = 108°$. Law of Sines: $\dfrac{BC}{\sin 108°} = \dfrac{100}{\sin 30°}$. Therefore, $BC = \dfrac{100 \sin 108°}{\sin 30°} = 190.2$. [6]

* **23.** **(E)** Plot the graph of $4 \sin x + 3$ in radian mode in a $[0, 2\pi]$ by $[-2, 8]$ window. Use CALC/zero twice to find the correct answer choice.

An alternative solution is to solve the equation for $\sin x$ and use your calculator, in radian mode: $\sin^{-1}\left(-\dfrac{3}{4}\right) = -0.848$.

Since this value is not between 0 and 2π, you must find the value of x in the required interval that has the same terminal side ($2\pi - 0.848 = 5.435$) as well as the third quadrant angle that has the same reference angle ($\pi + 0.848 = 3.990$). [7]

24. **(A)** The formula for the sum of a geometric series is $S = \dfrac{a_1}{1 - r}$, where a_1 is the first term and r is the common ratio. In this problem, $a_1 = 6$ and $r = \dfrac{4}{6} = \dfrac{2}{3}$. Therefore,

$$S = \dfrac{6}{1 - \dfrac{2}{3}} = 18. \text{ [19]}$$

* **25.** **(D)** Enter $\dfrac{1}{2 + 6i}$ into your calculator and press MATH/ENTER/ENTER/ to get FRACTIONAL real and imaginary parts.

An alternative solution is to multiply the numerator and the denominator of $\dfrac{1}{2 + 6i}$ by the complex conjugate $2 - 6i$:

$$\frac{1}{2 + 6i} \cdot \frac{2 - 6i}{2 - 6i} = \frac{2 - 6i}{4 - (-36)} = \frac{1}{20} - \frac{3}{20} i \text{ [17]}$$

* **26.** **(D)** The measure $\dfrac{3\pi}{14}$ of the central angle is superfluous. Areas of similar figures are proportional to the squares of linear measures associated with those figures. Since the ratio of the areas is $1 : 2$, the ratio of the radii is $1 : \sqrt{2}$, or approximately $0.71 : 1$. [7]

* **27.** **(B)** The line cuts the x-axis at 3 and the y-axis at 4 to form a right triangle that, when rotated about the y-axis, forms a cone with radius 3 and altitude 4.

Volume $= \dfrac{1}{3}\pi r^2 h = \dfrac{1}{3}\pi(9)(4) = 12\pi \approx 37.7$. [15]

* **28.**   **(B)** Since there are 4 broken transistors, there must be 8 good ones. P(first pick is good) $= \dfrac{8}{12}$. Of the remaining 11 transistors, 7 are good, and so P(second pick is good) $= \dfrac{7}{11}$. Finally, P(third pick is good) $= \dfrac{6}{10}$. Therefore:

$$P(\text{all three are good}) = \frac{8}{12} \cdot \frac{7}{11} \cdot \frac{6}{10} = \frac{14}{55} \approx 0.255$$

Alternative Solution: Note that there are $\dbinom{8}{3} = 56$ ways to select 3 good transistors.

There are $\dbinom{12}{3} = 220$ ways to select any 3 transistors.

$$P(\text{3 good ones}) = \frac{56}{220} = \frac{14}{55} \approx 0.255 \ [22]$$

29.   **(E)** The domain of the cube root function is all real numbers, so x can be any real number. [1]

30.   **(E)** Divide each term in the numerator and the denominator by x^4. As x increases or decreases, all but the first terms in the numerator and the denominator approach zero, leaving a ratio of $\dfrac{3}{2}$. [9]

* **31.**   **(C)** Plot the graphs of $Y_1 = 250e^{\frac{-t}{500}}$ and $Y_2 = 100$. Using the answer choices and the information in the problem, choose a [100,600] by [50,300] window, and find the point when Y_1 and Y_2 intersect. The solution is the x-coordinate of this point.

An alternative method is to substitute 250 for a and 100 for E to get $100 = 250e^{\frac{-t}{500}}$. Divide both sides by 250. Take the ln of both sides of the equation. Finally, multiply both sides by -500 to get $t = -500 \ln \dfrac{2}{5} \approx 458$. [8]

32.   **(A)** Using the Remainder Theorem, substitute -1 for x. Since $2n + 3$ and $2n + 1$ are odd, and $2n + 2$ is even, this substitution yields $3(-1) - 4(1) + 5(-1) - 8 = -20$. [4]

33.   **(A)** There are $4! = 24$ arrangements of all 4 men. If man A is on the one end of the line, there are $(2)3! = 12$ arrangements of the other 3 men. Therefore, the probability that man A is at an end seat is $\dfrac{12}{24} = \dfrac{1}{2}$. [22]

34.   **(C)** The summation indicates that 5 be summed 8 times (when $i = 3, 4, \ldots, 11$), so the sum is $8 \times 5 = 40$. [19]

* **35.**   **(E)** Graph $y = \sqrt[4]{3x^2 + 7}$ and $y = -\sqrt[4]{3x^2 + 7}$ in a standard window. Observe that the graph is symmetrical with respect to the x-axis, the y-axis, and the origin.

An alternative solution is to observe that if x is replaced by $-x$, if y is replaced by $-y$, or if both replacements take place, the equation is unaffected. Therefore, all three symmetries are present. [1]

36. **(C)** From the Venn diagram below you get the following equations:

All students

Fr   Sp

Take neither course

a b c d

(1) $a + b + c + d = 30$
(2) $b + c = 20$
(3) $c + d = 15$
$a = 5$

Subtract equation (2) from equation (1): $a + d = 10$. Since $a = 5$, $d = 5$. Substituting 5 for d in equation (3) leaves $c = 10$. [16]

37. **(D)** First note that $f(x + h) = (x + h)^2 = x^2 + 2xh + h^2$. So $f(x + h) - f(x) = 2xh + h^2$. Divide this expression by h to get the correct answer choice. [1]

38. **(A)** If $y = z = 0$, $x = 6$, so the plane intersects the x-axis at 6. By similar reasoning, the plane intersects the y-axis at 5 and the z-axis at 3. The figure below shows a pyramid with vertex $(0,0,3)$. The base is a right triangle with area $\frac{1}{2}(6)(5) = 15$. Therefore, the volume of the pyramid is $V = \frac{1}{3}(15)(3) = 15$. [15]

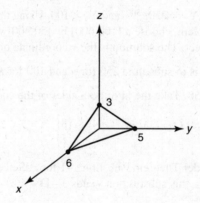

39.   **(D)** Replace $f(x)$ with y, interchange x and y, and solve for y. This will give a formula for the inverse of f.

$$x = \frac{1}{y-5} - 1$$

$$x + 1 = \frac{1}{y-5}$$

$$y - 5 = \frac{1}{x+1}$$

$$y = \frac{1}{x+1} + 5$$

The domain of f^{-1} is the range of f, and f^{-1} is defined for all real numbers except $x = -1$. [1]

40.   **(A)** The number of 2s must exceed the number of other values. Some of the choices can be eliminated. Choice B: one integer could be 2. Choice C: x or y could be 2, but not necessarily. Choices D and E: if $x = 3$ and $y = 1$, there will be no mode. Therefore, Choice A is the answer. [22]

✳ 41.   **(A)** Enter $\sqrt{2x+3}$ into Y_1, x^2 into Y_2, $Y_1(Y_2(X))$ into Y_3, and $Y_2(Y_1(X))$ into Y_4. De-select Y_1 and Y_2, and graph Y_3 and Y_4 in a $[-1.5,2]$ by $[0,5]$ window (because $2x + 3 \geq 0$). Use CALC/intersect to find the correct answer, Choice A.

An alternative solution is to evaluate $f(g(x)) = \sqrt{2x^3 + 3}$ and $g(f(x)) = 2x + 3$, set the two equal, square both sides, and solve the resulting quadratic:

$$2x^2 + 3 = (2x + 3)^2 = 4x^2 + 12x + 9 \text{ or } x^2 + 6x + 3 = 0$$

The Quadratic Formula yields $x = \dfrac{-6 + \sqrt{24}}{2}$ or $x = \dfrac{-6 - \sqrt{24}}{2}$. However, the second

is not in the domain of g, so $x = \dfrac{-6 + \sqrt{24}}{2} \approx -0.55$ is the only solution. [1]

42.   **(D)** Solve the first inequality to get $1 \leq x \leq 2$. Solve the second inequality to get $-2 \leq y \leq 1$. The smallest product xy possible is -4, and the largest product xy possible is $+2$. [4]

✳ 43.   **(C)** Use your calculator to evaluate $\sin\left(180° - \sin^{-1}\dfrac{1}{3} - \sin^{-1}\dfrac{1}{4}\right) \approx 0.56$. [6]

✳ 44.   **(A)** Graph $y = \dfrac{|x-1|}{x}$ and $y = 2$ in the standard window. There is a vertical asymptote

at $x = 0$, and the curve is above the horizontal $y = 2$ just to the right of 0. Choice A is the only possibility, but to be sure, use CALC/intersect to determine that the point of

intersection is at $x = \dfrac{1}{3}$.

An alternative solution is to use the associated equation $\dfrac{|x-1|}{x} = 2$ to find boundary values and then test points. Since the left side is undefined when $x = 0$, zero is one boundary value. Multiply both sides by x to get $|x - 1| = 2x$ and analyze the two cases for absolute value. If $x - 1 \geq 0$, then $x - 1 = 2x$ so $x = -1$, which is impossible because $x \geq 1$. Therefore, $x - 1 < 0$, so $1 - x = 2x$ or $x = \dfrac{1}{3}$ is the other boundary value. Testing points in the intervals $(-\infty, 0)$, $\left(0, \dfrac{1}{3}\right)$, and $\left(\dfrac{1}{3}, \infty\right)$ yields the correct answer, Choice A. [11]

* 45. **(D)** Translating f 2 units right and 3 units down results in $g(x) = -(x - 2)^3 - 1$. Then $g(-1.2) \approx 31.77$. [12]

46. **(D)** The volume of the prism is the area of the base times the height. The figure $EABC$ is a pyramid with base triangle ABC and height BE, the same as the base and height of the prism. The volume of the pyramid is $\dfrac{1}{3}$ (base times height), $\dfrac{1}{3}$ the volume of the prism. Therefore, the other solid, $EACFD$, is $\dfrac{2}{3}$ the volume of the prism. The ratio of the volumes is $\dfrac{1}{2}$. [15]

47. **(E)** For $x < 6$, the graph of f is a parabola whose vertex is at the point $(1, -2)$ with -2 as the minimum value. For $x \geq 6$, the value of $-2x + 5$ is less than or equal to $-2(6) + 5 = -7$. Therefore, $y \geq -2$ or $y \leq -7$. [1]

48. **(E)** Get the center axis form of the equation by completing the square:

$3x^2 - 6x + 2y^2 + 8y = 1$

$3(x^2 - 2x + 1) + 2(y^2 + 4y + 4) = 1 + 3 + 8 = 12$, which leads to

$\dfrac{3(x-1)^2}{12} + \dfrac{2(y+2)^2}{12} = 1$ and finally to $\dfrac{(x-1)^2}{4} + \dfrac{(y+2)^2}{6} = 1.$

Thus, half the major axis is $\sqrt{6}$, making the major axis $2\sqrt{6}$. [13]

* 49. **(D)** The probability that a student is a girl at this high school is 0.6. The probability that a student who plays a sport is a girl is 0.65. Therefore, the probability that a girl student plays a sport is $(0.6)(0.65) = 0.39$. [22]

* 50. **(A)** According to the factor theorem, substituting 7 for x yields 0. Therefore:

$$7^3 - 3(7)^2 k^3 - 13(7) - 7 = 0 \text{ , or } k = \sqrt[3]{\dfrac{245}{147}} \approx 1.19 \text{ [4]}$$

Self-Evaluation Chart for Practice Test 3

Subject Area	Questions and Review Section							Right	Number Wrong	Omitted
Algebra and Functions	1	2	3	4	7	10				
(24 questions)	2	4	1	1	3	9		____	____	____
	13	14	15	19	20	21				
	11	1	4	10	1	1		____	____	____
	29	30	31	32	35	37				
	1	9	8	4	1	1		____	____	____
	39	41	42	44	47	50				
	1	1	4	11	1	4		____	____	____
Trigonometry	5	8	12	22	23	26				
(7 questions)	7	7	7	6	7	7		____	____	____
	43									
	6							____	____	____
Coordinate and Three-	6	9	11	27	38	45				
Dimensional Geometry	15	12	15	15	15	12		____	____	____
(8 questions)										
	46	48								
	15	13						____	____	____
Number and Operations	16	17	18	24	25	34	36			
(7 questions)	20	16	19	19	17	19	16	____	____	____
Data Analysis, Statistics,	28	33	40	49						
and Probability	22	22	22	22						
(4 questions)								____	____	____
TOTALS								____	____	____

Evaluate Your Performance on Practice Test 3	
Rating	**Number Right**
Excellent	41–50
Very good	33–40
Above average	25–32
Average	15–24
Below average	Below 15

CALCULATING YOUR SCORE

Raw score R = number right $- \dfrac{1}{4}$ (number wrong), rounded = _____

Approximate scaled score $S = 800 - 10(44 - R)$ = _____

If $R \geq 44$, $S = 800$.

ANSWER SHEET
Practice Test 4

1. Ⓐ Ⓑ Ⓒ Ⓓ Ⓔ
2. Ⓐ Ⓑ Ⓒ Ⓓ Ⓔ
3. Ⓐ Ⓑ Ⓒ Ⓓ Ⓔ
4. Ⓐ Ⓑ Ⓒ Ⓓ Ⓔ
5. Ⓐ Ⓑ Ⓒ Ⓓ Ⓔ
6. Ⓐ Ⓑ Ⓒ Ⓓ Ⓔ
7. Ⓐ Ⓑ Ⓒ Ⓓ Ⓔ
8. Ⓐ Ⓑ Ⓒ Ⓓ Ⓔ
9. Ⓐ Ⓑ Ⓒ Ⓓ Ⓔ
10. Ⓐ Ⓑ Ⓒ Ⓓ Ⓔ
11. Ⓐ Ⓑ Ⓒ Ⓓ Ⓔ
12. Ⓐ Ⓑ Ⓒ Ⓓ Ⓔ
13. Ⓐ Ⓑ Ⓒ Ⓓ Ⓔ

14. Ⓐ Ⓑ Ⓒ Ⓓ Ⓔ
15. Ⓐ Ⓑ Ⓒ Ⓓ Ⓔ
16. Ⓐ Ⓑ Ⓒ Ⓓ Ⓔ
17. Ⓐ Ⓑ Ⓒ Ⓓ Ⓔ
18. Ⓐ Ⓑ Ⓒ Ⓓ Ⓔ
19. Ⓐ Ⓑ Ⓒ Ⓓ Ⓔ
20. Ⓐ Ⓑ Ⓒ Ⓓ Ⓔ
21. Ⓐ Ⓑ Ⓒ Ⓓ Ⓔ
22. Ⓐ Ⓑ Ⓒ Ⓓ Ⓔ
23. Ⓐ Ⓑ Ⓒ Ⓓ Ⓔ
24. Ⓐ Ⓑ Ⓒ Ⓓ Ⓔ
25. Ⓐ Ⓑ Ⓒ Ⓓ Ⓔ
26. Ⓐ Ⓑ Ⓒ Ⓓ Ⓔ

27. Ⓐ Ⓑ Ⓒ Ⓓ Ⓔ
28. Ⓐ Ⓑ Ⓒ Ⓓ Ⓔ
29. Ⓐ Ⓑ Ⓒ Ⓓ Ⓔ
30. Ⓐ Ⓑ Ⓒ Ⓓ Ⓔ
31. Ⓐ Ⓑ Ⓒ Ⓓ Ⓔ
32. Ⓐ Ⓑ Ⓒ Ⓓ Ⓔ
33. Ⓐ Ⓑ Ⓒ Ⓓ Ⓔ
34. Ⓐ Ⓑ Ⓒ Ⓓ Ⓔ
35. Ⓐ Ⓑ Ⓒ Ⓓ Ⓔ
36. Ⓐ Ⓑ Ⓒ Ⓓ Ⓔ
37. Ⓐ Ⓑ Ⓒ Ⓓ Ⓔ
38. Ⓐ Ⓑ Ⓒ Ⓓ Ⓔ
39. Ⓐ Ⓑ Ⓒ Ⓓ Ⓔ

40. Ⓐ Ⓑ Ⓒ Ⓓ Ⓔ
41. Ⓐ Ⓑ Ⓒ Ⓓ Ⓔ
42. Ⓐ Ⓑ Ⓒ Ⓓ Ⓔ
43. Ⓐ Ⓑ Ⓒ Ⓓ Ⓔ
44. Ⓐ Ⓑ Ⓒ Ⓓ Ⓔ
45. Ⓐ Ⓑ Ⓒ Ⓓ Ⓔ
46. Ⓐ Ⓑ Ⓒ Ⓓ Ⓔ
47. Ⓐ Ⓑ Ⓒ Ⓓ Ⓔ
48. Ⓐ Ⓑ Ⓒ Ⓓ Ⓔ
49. Ⓐ Ⓑ Ⓒ Ⓓ Ⓔ
50. Ⓐ Ⓑ Ⓒ Ⓓ Ⓔ

Practice Test 4

T ear out the preceding answer sheet. Decide which is the best choice by rounding your answer when appropriate. Blacken the corresponding space on the answer sheet. When finished, check your answers with those at the end of the test. For questions that you got wrong, note the chapters containing the material that you must review. Also, if you do not fully understand how you arrived at some of the correct answers, you should review the appropriate chapters. Finally, fill out the Self-Evaluation Chart on page 297 in order to pinpoint the topics that gave you the most difficulty.

50 questions: 1 hour

Directions: Decide which answer choice is best. If the exact numerical value is not one of the answer choices, select the closest approximation. Fill in the circle on the answer sheet that corresponds to your choice.

Notes:
(1) You will need to use a scientific or graphing calculator to answer some of the questions.
(2) You will have to decide whether to put your calculator in degree or radian mode for some problems.
(3) All figures that accompany problems are plane figures unless otherwise stated. Figures are drawn as accurately as possible to provide useful information for solving the problem, except when it is stated in a particular problem that the figure is not drawn to scale.
(4) Unless otherwise indicated, the domain of a function is the set of all real numbers for which the functional value is also a real number.

Reference Information. The following formulas are provided for your information.

Volume of a right circular cone with radius r and height h: $V = \dfrac{1}{3}\pi r^2 h$

Volume of a sphere of radius r: $V = \dfrac{4}{3}\pi r^3$

Surface area of a sphere of radius r: $S = 4\pi r^2$

Volume of a pyramid of base area B and height h: $V = \dfrac{1}{3}Bh$

1. If point (a,b) lies on the graph of function f, which of the following points must lie on the graph of the inverse f?

(A) (a,b)
(B) $(-a,b)$
(C) $(a,-b)$
(D) (b,a)
(E) $(-b,-a)$

2. Harry had grades of 70, 80, 85, and 80 on his quizzes. If all quizzes have the same weight, what grade must he get on his next quiz so that his average will be 80?

(A) 85
(B) 90
(C) 95
(D) 100
(E) more than 100

3. If L is a line through $(-1,3)$ and $(4,2)$, for what value of k is $(3,k)$ on L?

(A) -2
(B) 0
(C) 2.2
(D) 2.5
(E) 2.8

4. If $\log_b x = p$ and $\log_b y = q$, then $\log_b xy =$

(A) pq
(B) $p + q$
(C) $\dfrac{p}{q}$
(D) $p - q$
(E) p^q

5. The sum of the roots of $3x^3 + 4x^2 - 4x = 0$ is

(A) $-\dfrac{4}{3}$
(B) $-\dfrac{3}{4}$
(C) 0
(D) $\dfrac{4}{3}$
(E) 4

PRACTICE TEST 4

6. If $f(x) = x - \dfrac{1}{x}$, then $f(a) + f\left(\dfrac{1}{a}\right) =$

    (A) 0

    (B) $2a - \dfrac{2}{a}$

    (C) $a - \dfrac{1}{a}$

    (D) $\dfrac{a^4 - a^2 + 1}{a(a^2 - 1)}$

    (E) 1

7. If $f(x) = \log(x + 1)$, what is $f^{-1}(3)$?

    (A) 0.60
    (B) 4
    (C) 999
    (D) 1,001
    (E) 10,000

8. If $f(x) \geq 0$ for all x, then $f(2 - x)$ is

    (A) ≥ -2
    (B) ≥ 0
    (C) ≥ 2
    (D) ≤ 0
    (E) ≤ 2

9. How many four-digit numbers can be formed from the numbers 0, 2, 4, 8 if no digit is repeated?

    (A) 18
    (B) 24
    (C) 27
    (D) 36
    (E) 64

10. If $x - 1$ is a factor of $x^2 + ax - 4$, then a has the value

    (A) 4
    (B) 3
    (C) 2
    (D) 1
    (E) none of the above

11. If 10 coins are to be flipped and the first 5 all come up heads, what is the probability that exactly 3 more heads will be flipped?

    (A) 0.0439
    (B) 0.1172
    (C) 0.1250
    (D) 0.3125
    (E) 0.6000

12. If $i = \sqrt{-1}$ and n is a positive integer, which of the following statements is FALSE?

    (A) $i^{4n} = 1$
    (B) $i^{4n+1} = -i$
    (C) $i^{4n+2} = -1$
    (D) $i^{n+4} = i^n$
    (E) $i^{4n+3} = -i$

13. If $\log_r 3 = 7.1$, then $\log_r \sqrt{3} =$

    (A) 2.66

    (B) 3.55

    (C) $\dfrac{\sqrt{3}}{r}$

    (D) $\dfrac{7.1}{r}$

    (E) $\sqrt[r]{7.1}$

14. If $f(x) = 4x^2$ and $g(x) = f(\sin x) + f(\cos x)$, then $g(23°)$ is

    (A) 1
    (B) 4
    (C) 4.29
    (D) 5.37
    (E) 8

15. What is the sum of the roots of the equation

$$(x - \sqrt{2})(x^2 - \sqrt{3}x + \pi) = 0?$$

    (A) −0.315
    (B) −0.318
    (C) 1.414
    (D) 3.15
    (E) 4.56

16. Which of the following equations has (have) graphs that consist of two perpendicular lines?

    I. $xy = 0$
    II. $|y| = |x|$
    III. $|xy| = 1$

    (A) only I
    (B) only II
    (C) only III
    (D) only I and II
    (E) I, II, and III

17. If $f(x) = (4 - x)^2$ and $g(x) = \sqrt{x}$, then $(g \circ f)(x) =$

(A) $|4 - x|$

(B) $|2 - x|$

(C) $\sqrt{x(4 - x)^2}$

(D) $\left(\sqrt{4 - x}\right)^2$

(E) $x - 4$

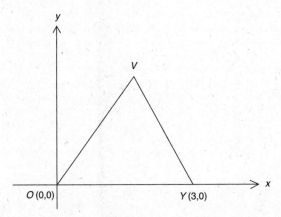

18. In the figure above, if $VO = VY$, what is the slope of segment VO?

(A) $-\sqrt{3}$

(B) $-\dfrac{\sqrt{3}}{2}$

(C) $\dfrac{\sqrt{3}}{2}$

(D) $\sqrt{3}$

(E) cannot be determined from the given information

19. A cylindrical bar of metal has a base radius of 2 and a height of 9. It is melted down and reformed into a cube. A side of the cube is

(A) 2.32

(B) 3.84

(C) 4.84

(D) 97.21

(E) 113.10

20. The graph of $y = (x + 2)(2x - 3)$ can be expressed as a set of parametric equations. If $x = 2t - 2$ and $y = f(t)$, then $f(t) =$

(A) $2t(4t - 5)$

(B) $(2t - 2)(4t - 7)$

(C) $2t(4t - 7)$

(D) $(2t - 2)(4t - 5)$

(E) $2t(4t + 1)$

21. The inverse of $y = \dfrac{2x+1}{3-x}$ is

(A) $y = \dfrac{-2x-1}{x-3}$

(B) $y = \dfrac{3-x}{2x+1}$

(C) $y = \dfrac{3x+1}{x-2}$

(D) $y = \dfrac{3x-1}{x+2}$

(E) $y = \dfrac{3x+1}{2-x}$

22. Rent-a-Rek has 27 cars available for rental. Twenty of these are compact, and 7 are midsize. If two cars are selected at random, what is the probability that both are compact?

(A) 0.0576
(B) 0.0598
(C) 0.481
(D) 0.521
(E) 0.541

23. If a and b are real numbers, with $a > b$ and $|a| < |b|$, then

(A) $a > 0$
(B) $a < 0$
(C) $b > 0$
(D) $b < 0$
(E) none of the above

24. If $[x]$ is defined to represent the greatest integer less than or equal to x, and $f(x) = \left| x - [x] - \dfrac{1}{2} \right|$, the maximum value of $f(x)$ is

(A) -1
(B) 0
(C) $\dfrac{1}{2}$
(D) 1
(E) 2

PRACTICE TEST 4

25. As x gets closer to 2, $\dfrac{x^3 - 8}{x^2 - 4}$ approaches

(A) 0
(B) 1
(C) 2
(D) 3
(E) ∞

26. A right circular cone whose base radius is 4 is inscribed in a sphere of radius 5. What is the ratio of the volume of the cone to the volume of the sphere?

(A) 0.222 : 1
(B) 0.256 : 1
(C) 0.288 : 1
(D) 0.333 : 1
(E) 0.864 : 1

27. If $x_0 = 1$ and $x_{n+1} = \sqrt[3]{2x_n}$, then $x_3 =$

(A) 1.260
(B) 1.361
(C) 1.396
(D) 1.408
(E) 1.412

28. The y-intercept of $y = \left| \sqrt{2} \csc 3\left(x + \dfrac{\pi}{5} \right) \right|$ is

(A) 0.22
(B) 0.67
(C) 1.41
(D) 1.49
(E) 4.58

29. If the center of the circle $x^2 + y^2 + ax + by + 2 = 0$ is point $(4, -8)$, then $a + b =$

(A) -8
(B) -4
(C) 4
(D) 8
(E) 24

30. If $p(x) = 3x^2 + 9x + 7$ and $p(a) = 2$, then $a =$

(A) only 0.736
(B) only -2.264
(C) 0.736 or 2.264
(D) 0.736 or -2.264
(E) -0.736 or -2.264

31. If 2 is a root of the equation $x^4 + 2x^3 - 3x^2 + kx - 4 = 0$, then $k =$

(A) −12
(B) −10
(C) −8
(D) 4
(E) 8

32. If $\sin A = \dfrac{3}{5}$, $90° \leq A \leq 180°$, then $\sin 2A =$

(A) $-\dfrac{6}{5}$

(B) $-\dfrac{24}{25}$

(C) 0

(D) $\dfrac{24}{25}$

(E) $\dfrac{6}{5}$

33. A family has three children. Assuming that the probability of having a boy is 0.5, what is the probability that at least one child is a boy?

(A) 0.875
(B) 0.67
(C) 0.5
(D) 0.375
(E) 0.25

34. If $\sec 1.4 = x$, find the value of $\csc(2 \tan^{-1} x)$.

(A) 0.33
(B) 0.87
(C) 1.00
(D) 1.06
(E) 3.03

35. The graph of $|y - 1| = |x + 1|$ forms an X. The two branches of the X intersect at a point whose coordinates are

(A) (1,1)
(B) (−1,1)
(C) (1,−1)
(D) (−1,−1)
(E) (0,0)

36. For what value of x between 0° and 360° does $\cos 2x = 2 \cos x$?

(A) 68.5° or 291.5°
(B) only 68.5°
(C) 103.9° or 256.1°
(D) 90° or 270°
(E) 111.5° or 248.5°

PRACTICE TEST 4

37. A data set has a mean of 12 and a standard deviation of 2. If 5 is added to twice each data value, the new data set has

   (A) a mean of 24 and a standard deviation of 10
   (B) a mean of 24 and a standard deviation of 15
   (C) a mean of 29 and a standard deviation of 15
   (D) a mean of 29 and a standard deviation of 4
   (E) a mean of 29 and a standard deviation of 10

38. For each positive integer n, let S_n = the sum of all positive integers less than or equal to n. Then S_{51} equals

   (A) 50
   (B) 51
   (C) 1,250
   (D) 1,275
   (E) 1,326

39. If the graphs of $3x^2 + 4y^2 - 6x + 8y - 5 = 0$ and $(x - 2)^2 = 4(y + 2)$ are drawn on the same coordinate system, at how many points do they intersect?

   (A) 0
   (B) 1
   (C) 2
   (D) 3
   (E) 4

40. If $\log_x 2 = \log_3 x$ is satisfied by two values of x, what is their sum?

   (A) 0
   (B) 1.73
   (C) 2.35
   (D) 2.81
   (E) 3.14

41. Which of the following lines are asymptotes for the graph of $y = \dfrac{3x^2 - 13x - 10}{x^2 - 4x - 5}$?

   I. $x = -1$
   II. $x = 5$
   III. $y = 3$

   (A) I only
   (B) II only
   (C) I and II
   (D) I and III
   (E) I, II, and III

42. If $\dfrac{3 \sin 2\theta}{1 - \cos 2\theta} = \dfrac{1}{2}$ and $0° \le \theta \le 180°$, then $\theta =$

(A) 0°
(B) 0° or 180°
(C) 80.5°
(D) 0° or 80.5°
(E) 99.5°

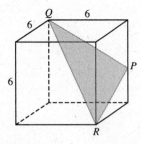

43. The cube in the figure above has edges of length 6. If point P is the midpoint of an edge, what is the perimeter of $\triangle QPR$?

(A) 13.41
(B) 14.28
(C) 17.10
(D) 21.90
(E) 23.81

44. Two positive numbers, a and b, are in the sequence 4, a, b, 12. The first three numbers form a geometric sequence, and the last three numbers form an arithmetic sequence. The difference $b - a$ equals

(A) 1
(B) $1\dfrac{1}{2}$
(C) 2
(D) $2\dfrac{1}{2}$
(E) 3

45. A sector of a circle has an arc length of 2.4 feet and an area of 14.3 square feet. How many degrees are in the central angle?

(A) 63.4°
(B) 20.2°
(C) 14.3°
(D) 12.9°
(E) 11.5°

46. The y-coordinate of one focus of the ellipse $36x^2 + 25y^2 + 144x - 50y - 731 = 0$ is

(A) −2
(B) 1
(C) 3.32
(D) 4.32
(E) 7.81

PRACTICE TEST 4

USE THIS SPACE FOR SCRATCH WORK

47. In the figure above, *ABCD* is a square. *M* is the point one-third of the way from *B* to *C*. *N* is the point one-half of the way from *D* to *C*. Then θ =

(A) 50.8°
(B) 45.0°
(C) 36.9°
(D) 36.1°
(E) 30.0°

48. If *f* is a linear function such that *f*(7) = 5, *f*(12) = –6, and *f*(*x*) = 23.7, what is the value of *x*?

(A) –3.2
(B) –1.5
(C) 1
(D) 2.4
(E) 3.1

49. In January 1980, the population of a certain city was 900,000. Assuming an annual growth rate of 3.7%, the population of that city (in millions) for *t* years after 1980 can be modeled by the function *P* = 0.9(1.037)^*t*. According to the model, the population growth from January 1990 through January 2000 was

(A) 485,000
(B) 533,000
(C) 567,000
(D) 885,000
(E) 1,294,000

50. In a group of 12 people, 2 have green eyes. Two people are selected at random from the group. What is the probability that neither has green eyes?

(A) 0.05
(B) 0.31
(C) 0.50
(D) 0.68
(E) 0.76

If there is still time remaining, you may review your answers.

PRACTICE TEST 4   289

1. D	14. B	27. C	40. D
2. A	15. D	28. D	41. D
3. C	16. D	29. D	42. C
4. B	17. A	30. E	43. E
5. A	18. E	31. C	44. E
6. A	19. C	32. B	45. E
7. C	20. C	33. A	46. D
8. B	21. D	34. E	47. B
9. A	22. E	35. B	48. B
10. B	23. D	36. E	49. C
11. D	24. C	37. D	50. D
12. B	25. D	38. E	
13. B	26. B	39. C	

ANSWERS EXPLAINED

The following explanations are keyed to the review portions of this book. The number in brackets after each explanation indicates the appropriate chapter in the Review of Major Topics (Part 2).

An asterisk appears next to those solutions for which a graphing calculator is necessary or helpful.

1. **(D)** Since inverse functions are symmetric about the line $y = x$, if point (a,b) lies on f, point (b,a) must lie on f^{-1}. [1]

* 2. **(A)** Average $= \dfrac{70 + 80 + 85 + 80 + x}{5} = 80$. Therefore, $x = 85$. [21]

3. **(C)** The slope of the line is $\dfrac{2-3}{4-(-1)} = -\dfrac{1}{5}$. Therefore, $\dfrac{k-2}{3-4} = -\dfrac{1}{5}$. Cross-multiplying

    yields $5k - 10 = 1$, so $k = \dfrac{11}{5} = 2.2$. [2]

4. **(B)** The bases are the same, so the log of a product equals the sum of the logs. [8]

* 5. **(A)** Plot the graph of $y = 3x^3 + 4x^2 - 4x$ in the standard window and use CALC/zero to find the three zeros of this function. Sum these three values to get the correct answer choice.

    An alternative solution is to first observe that x factors out, so that $x = 0$ is one zero.

    The other factor is a quadratic, so the sum of its zeros is $-\dfrac{b}{a} = -\dfrac{4}{3}$. [4]

6. **(A)** $f(a) = a - \dfrac{1}{a}$ and $f\left(\dfrac{1}{a}\right) = \dfrac{1}{a} - a$. Therefore, $f(a) + f\left(\dfrac{1}{a}\right) = 0$. [1]

7. **(C)** By the definition of f, $\log(x + 1) = 3$. Therefore, $x + 1 = 10^3 = 1{,}000$ and $x = 999$. [8]

8. **(B)** The $f(2 - x)$ just shifts and reflects the graph horizontally; it does not have any vertical effect on the graph. Therefore, regardless of what is substituted for x, $f(x) \geq 0$. [12]

9. **(A)** Only 3 of the numbers can be used in the thousands place, 3 are left for the hundreds place, 2 for the tens place, and only one for the units place. $3 \cdot 3 \cdot 2 \cdot 1 = 18$. [16]

10. **(B)** Substituting 1 for x gives $1 + a - 4 = 0$, and so $a = 3$. [4]

* **11.**   **(D)** The first 5 flips have no effect on the next 5 flips, so the problem becomes "What is the probability of getting exactly 3 heads in 5 flips of a coin?" $\binom{5}{3} = 10$ outcomes

contain 3 heads out of a total of $2^5 = 32$ possible outcomes. $P(3\text{H}) = \dfrac{5}{16} \approx 0.3125$. [22]

12.   **(B)** Powers of i repeat in groups of 4: $i^0 = 1$, $i^1 = i$, $i^2 = -1$, $i^3 = -i$. Then $i^4 = 1$, and so forth. Therefore, $i^{4n} = i^4 = 1$, so Choice A is true. $i^{4n+2} = i^2 = -1$, so Choice C is true. $i^{n+4} = i^n$ because the powers repeat in groups of 4, so Choice D is true. $i^{4n+3} = i^3 = -i$, so Choice E is true. However, $i^{4n+1} = i^1 = i$, so Choice B is *not* true. [17]

13.   **(B)** Since $\sqrt{x} = x^{1/2}$, $\log_r \sqrt{3} = \log_r 3^{1/2} = \dfrac{1}{2} \log_r 3 = 3.55$. [8]

14.   **(B)** The 23° is superfluous because $g(x) = f(\sin x) + f(\cos x) = 4 \sin^2 x + 4 \cos^2 x = 4(\sin^2 x + \cos^2 x) = 4(1) = 4$. [7]

* **15.**   **(D)** This is a tricky problem. If you just plot the graph of $y = \left(x - \sqrt{2}\right)\left(x^2 - \sqrt{3}x + \pi\right)$,

you will see only one real zero, at approximately 1.414 ($\approx \sqrt{2}$). This is because the zeros of the quadratic factor are imaginary. Since, however, they are imaginary conjugates, their sum is real—namely twice the real part. Therefore, graphing the function, using CALC/zero to find the zeros, and summing them will give you the wrong answer.

To get the correct answer, you must use the fact that the sum of the zeros of the

quadratic factor is $-\dfrac{b}{a} = \sqrt{3} \approx 1.732$. Since the zero of the linear factor is $\sqrt{2} \approx 1.414$,

the sum of the zeros is about $1.732 + 1.414 \approx 3.15$. [4]

16.   **(D)** Graph I consists of the lines $x = 0$ and $y = 0$, which are the coordinate axes and are therefore perpendicular. Graph II consists of $y = |x|$ and $y = -|x|$, which are at $\pm 45°$ to the coordinate axes and are therefore perpendicular. Graph III consists of the hyperbolas $xy = 1$ and $xy = -1$. Therefore, the correct answer choice is Choice D.

There are two reasons why a graphing calculator solution is not recommended here. One is that equations, not functions, are given, and solving these equations so that they can be graphed involves two branches each. The other reason is that even with graphs, you would have to make judgments about perpendicularity. At a minimum, this would require you to graph the equations in a square window. [11]

17.   **(A)** $(g \circ f)(x) = g(f(x)) = \sqrt{(4-x)^2} = |4 - x|$. [1]

18.   **(E)** Since the y-coordinate of the point V could be at any height, the slope of VO could be any value. [2]

* **19.**   **(C)** Volume of cylinder $= \pi r^2 h = 36\pi = $ volume of cube $= s^3$. Therefore:

$$s = \sqrt[3]{36\pi} \approx 4.84 \ [15]$$

20. **(C)** Substitute $2t - 2$ for x. [10]

21. **(D)** Interchange x and y, and solve for y:

$$x = \frac{2y+1}{3-y}$$
$$3x - xy = 2y + 1$$
$$3x - 1 = 2y + xy$$
$$y(2 + x) = 3x - 1$$
$$y = \frac{3x-1}{x+2} \quad [1]$$

✳ 22. **(E)** The probability that the first car selected is compact is $\frac{20}{27}$. There are 26 cars left, of which 19 are compact. The probability that the second car is also compact is:

$$\left(\frac{20}{27}\right) \cdot \left(\frac{19}{26}\right) \approx 0.541 \ [22]$$

23. **(D)** Here, a could be either positive or negative. However, b must be negative. [11]

✳ 24. **(C)** Plot the graph of $y = \mathrm{abs}(x - \mathrm{int}(x) - 1/2)$ in a $[-5,5]$ by $[-2,2]$ window and observe that the maximum value is $\frac{1}{2}$.

An alternative solution is to sketch a portion of the graph by hand and observe the maximum value. [11]

✳ 25. **(D)** Plot the graph $y = (x^3 - 8)(x^2 - 4)$ in the standard window. Using CALC/value, observe that y is not defined when $x = 2$. Therefore, enter 1.999 for x and observe that y is approximately equal to 3.

An alternative solution is to factor the numerator and the denominator, divide out $x - 2$, and substitute the limiting value 2 into the resulting expression:

$$\lim_{x \to 2} \frac{x^3 - 8}{x^2 - 4} = \lim_{x \to 2} \frac{(x-2)(x^2 + 2x + 2)}{(x-2)(x+2)}$$
$$= \lim_{x \to 2} \frac{(x^2 + 2x + 4)}{x + 2} = \frac{4 + 4 + 4}{2 + 2} = 3 \ [9]$$

✳ 26. **(B)** The sketch below will help you see that the height of the cone is $5 + 3 = 8$.

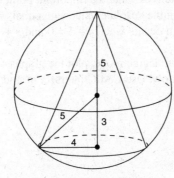

The volume of the cone is $V_c = \frac{1}{3}\pi r^2 h = \frac{1}{3}\pi(16)(8)$, and the volume of the sphere is

$V_s = \frac{4}{3}\pi r^3 = \frac{4}{3}\pi(125)$. The desired ratio is $V_c : V_s = 0.256 : 1$. [15]

* 27. **(C)** Enter 1 into your calculator. Then enter $\sqrt[3]{2\text{Ans}}$ three times, to accomplish three iterations that result in x_3 to get the correct answer, Choice C.

An alternative solution is to use the formula to evaluate $x_1, x_2,$ and x_3, in turn. [19]

* 28. **(D)** With your calculator in radian mode, plot the graph of $y = \text{abs}\left(\sqrt{2}\left(1/\sin\left(x + \pi/5\right)\right)\right)$ in a $[-1,1]$ by $[0,5]$ window. Use VALUE/X = 0 to determine that the y-intercept is approximately 1.49. [7]

29. **(D)** The equation of the circle is $(x - 4)^2 + (y + 8)^2 = r^2$. Multiplying out indicates that $a = -8$ and $b = 16$, and so $a + b = 8$. [13]

30. **(E)** Since $p(a) = 2, 3a^2 + 9a + 5 = 0$. Solve by using the Quadratic Formula to get the correct answer choice. [3]

31. **(C)** Since 2 is a root, the result should be 0 when you substitute 2 for x:
$16 + 16 - 12 + 2k - 4 = 0$. Solve this equation to get $k = -8$. [4]

32. **(B)** The formula for the sine of double an angle A is $\sin 2A = 2\sin A \cos A$. Since $\sin A = \frac{3}{5}$ and A is in the second quadrant, $\cos A = -\frac{4}{5}$. Substituting these values into the formula yields the answer, $-\frac{24}{25}$. [7]

* 33. **(A)** The probability that at least one is a boy is (1 – the probability that all 3 are girls) or $1 - (0.5)^3 = 0.875$. [22]

* 34. **(E)** Since $\sec 1.4 = x$, $\cos 1.4 = \frac{1}{x}$ or $x = \frac{1}{\cos 1.4}$. Therefore:

$$\csc(2\tan^{-1} x) = 1/\sin(2\tan^{-1}(1/\cos 1.4)) \approx 3.03 \ [7]$$

* 35. **(B)** The equation $|y - 1| = |x + 1|$ defines two functions: $y = \pm|x + 1| + 1$. Plot these graphs in the standard window and observe that they intersect at $(-1,1)$.

An alternative solution is to recall that the important point of an absolute value occurs when the expression within the absolute value sign equals zero. The important point of this absolute value problem occurs when $y - 1 = 0$ and $x + 1 = 0$, i.e., at $(-1,1)$. [11]

* 36. **(E)** With your calculator in degree mode, plot the graphs of $y = \cos 2x$ and $y = 2\cos x$ in a $[0,360]$ by $[-2,2]$ window. Use CALC/intersect to find the correct answer, Choice E. [7]

37. **(D)** The new mean is 5 more than twice the old mean, or $2(12) + 5 = 29$, but the new standard deviation is only twice the old one because adding a value to each data point does not affect the standard deviation. [21]

* **38.** **(E)** Enter LIST/MATH/sum(LIST/OPS/seq(X,X,1,51)) to compute the desired sum.

An alternative solution is to observe that the sequence is arithmetic with $t_1 = 1$ and $d = 1$. Use the formula for the sum of the first n terms of an arithmetic sequence:

$$S_{51} = \frac{51}{2}(2 + 50 \cdot 1) = 51 \cdot 26 = 1{,}326 \text{ [19]}$$

* **39.** **(C)** Complete the square in the first equation to get $3(x - 1)^2 + 4(y + 1)^2 = 12$. Solving this equation for y yields $y = \pm\sqrt{\dfrac{12 - 3(x-1)^2}{4}} - 1$. Solving for y in the second equation, $y = \dfrac{(x-2)^2}{4} - 2$. Plot the graphs of these three equations in the standard window to see that the graphs intersect in two places.

An alternative solution can be found by completing the square in the first equation and dividing by 12 to get the standard form equation of an ellipse, $\dfrac{(x-1)^2}{4} + \dfrac{(y+1)^2}{3} = 1$.

The second equation is the standard form of a parabola. Sketch these two equations and observe the number of points of intersection. [13]

* **40.** **(D)** Let $y = \log_x 2 = \log_3 x$. Converting to exponential form gives $x^y = 2$ and $3^y = x$. Substitute to get $3^{y^2} = 2$, which can be converted into $y^2 = \dfrac{\log 2}{\log 3} \approx 0.6309$. Thus, $y \approx \pm 0.7943$. This means that $3^{0.7943} = x \approx 2.393$ or $3^{-0.7943} = x \approx 0.4178$. Therefore, the sum of two xs is 2.81. [8]

41. **(D)** Factor the numerator and the denominator: $\dfrac{(3x+2)(x-5)}{(x+1)(x-5)}$. Since $(x - 5)$ divides out, the only vertical asymptote is at $x = -1$. Since the degree of the numerator and the denominator are equal, y approaches 3 as x approaches $\pm\infty$, so $y = 3$ is a horizontal asymptote. [9]

* **42.** **(C)** With your calculator in degree mode, plot the graphs of $y = 1/2$ and $y = (3 \sin(2x))/(1 - \cos(2x))$ in the window [0,180] by [−2,2]. Use CALC/intersect to find the one point of intersection in the specified interval at 80.5°.

An alternative solution is to cross-multiply the original equation and use the double angle formulas for sine and cosine, to get:

$$6 \sin 2\theta = 1 - \cos 2\theta$$
$$12 \sin\theta \cos\theta = 1 - (1 - 2\sin^2\theta) = 2\sin^2\theta$$
$$6 \sin\theta \cos\theta - \sin^2\theta = 0$$
$$\sin\theta(6 \cos\theta - \sin\theta) = 0$$

Therefore, $\sin\theta = 0$ or $\tan\theta = 6$. It follows that $\theta = 0°$, $180°$, or $80.5°$. The first two solutions make the denominator of the original equation equal to zero, so $80.5°$ is the only solution. [7]

43. **(E)** By the Pythagorean theorem, QP and RP each equal $\sqrt{6^2 + 3^2} = \sqrt{45}$. Apply the Pythagorean theorem to find $SR^2 = 6^2 + 6^2$, and $QR^2 = 6^2 + 6^2 + 6^2$, so $QR = 6\sqrt{3}$. Therefore, the perimeter of $\triangle QPR = 2\sqrt{45} + 6\sqrt{3} = 23.81$. [15]

44. **(E)** From the geometric sequence, $b = a\left(\dfrac{a}{4}\right)$. From the arithmetic sequence, $b - a = 12 - b$, or $2b - a = 12$. Substitute to get $2\left(\dfrac{a^2}{4}\right) - a = 12$. Then solve to find that $a = 6$ or -4. Eliminate -4 since a is given to be positive. Substituting the 6 gives $2b - 6 = 12$, which means that $b = 9$. Therefore, $b - a = 3$. [19]

45. **(E)** Since $s = r\theta$, then $2.4 = r\theta$, which implies that $r = \dfrac{2.4}{\theta}$. $A = \dfrac{1}{2}r^2\theta$, and so $14.3 = \dfrac{1}{2}r^2\theta$, which means that $r^2 = \dfrac{28.6}{\theta}$. Then $\left(\dfrac{2.4}{\theta}\right)^2 = \dfrac{28.6}{\theta}$, which implies that $2.4^2 = 28.6\theta$. Therefore, $\theta = \dfrac{5.76}{28.6} \approx 0.2014$ radians $\approx 11.5°$. [7]

46. **(D)** Complete the square and put the equation of the ellipse in standard form:

$$36x^2 + 25y^2 + 144x - 50y - 731 = 36(x+2)^2 + 25(y-1)^2 - 900$$

$$\frac{(x+2)^2}{25} + \frac{(y-1)^2}{36} = 1$$

The center of the ellipse is at $(-2,1)$, with $a^2 = 36$ and $b^2 = 25$, and the major axis is parallel to the y-axis. Each focus is $c = \sqrt{a^2 - b^2}$ units above and below the center. Therefore, the y-coordinates of the foci are $1 \pm \sqrt{11} \approx 4.32$ and -2.32. [13]

47. **(B)** Because you are bisecting one side and trisecting another side, it is convenient to let the lengths of the sides be a number divisible by both 2 and 3. Let $AB = AD = 6$. Thus $BM = 2$, $MC = 4$, and $CN = ND = 3$. Let $\angle NAD = x$, so that using right triangle NAD, $\tan x = \dfrac{3}{6} = 0.5$, which implies that $x = \tan^{-1} 0.5 \approx 26.6°$. Let $\angle MAB = y$, so that using right triangle MAB, $\tan y = \dfrac{2}{6}$, which implies that $y = \tan^{-1} \dfrac{1}{3} \approx 18.4°$. Therefore, $\theta \approx 90° - 26.6° - 18.4° \approx 45°$. [6]

48. **(B)** The slope of the line is $\dfrac{-6-5}{12-7} = -\dfrac{11}{5}$. An equation of the line is therefore $y - 5 = -\dfrac{11}{5}(x - 7)$. Substitute 23.7 for y and solve for x to get $x = -1.5$. [1]

49. **(C)** The population in January 1990 corresponds to $t = 10$, and the population in January 2000 corresponds to $t = 20$. Therefore, the population growth during this period is $0.9(1.037)^{20} - 0.9(1.037)^{10} \approx 0.567$ million $= 567,000$. [8]

50. **(D)** The probability that the first person selected does not have green eyes is $\dfrac{10}{12}$. Given that the first person selected does not have green eyes, the probability that the second person selected does not have green eyes is $\dfrac{9}{11}$, since there are eleven people left of which 9 do not have green eyes. Therefore, the probability that the first two people selected do not have green eyes is $\dfrac{10}{12} \times \dfrac{9}{11} \approx 0.68$. [22]

Self-Evaluation Chart for Practice Test 4

Subject Area	Questions and Review Section							Right	Number Wrong	Omitted
Algebra and Functions (25 questions)	1	3	4	5	6	7	10			
	1	2	8	4	1	8	4	____	____	____
	13	15	16	17	18	20	21			
	8	4	11	1	2	10	1	____	____	____
	23	24	25	30	31	35				
	11	11	9	3	4	11		____	____	____
	40	41	43	48	49					
	8	9	15	1	8			____	____	____
Trigonometry (8 questions)	14	28	32	34	36					
	7	7	7	7	7			____	____	____
	42	45	47							
	7	7	6					____	____	____
Coordinate and Three-Dimensional Geometry (6 questions)	8	19	26	29	39	46				
	12	15	15	13	13	13		____	____	____
Number and Operations (5 questions)	9	12	27	38	44					
	16	17	19	19	19			____	____	____
Data Analysis, Statistics, and Probability (6 questions)	2	11	22	33	37	50				
	21	22	22	22	21	22		____	____	____
TOTALS								____	____	____

Evaluate Your Performance on Practice Test 4	
Rating	**Number Right**
Excellent	41–50
Very good	33–40
Above average	25–32
Average	15–24
Below average	Below 15

CALCULATING YOUR SCORE

Raw score R = number right $- \frac{1}{4}$(number wrong), rounded = _____

Approximate scaled score $S = 800 - 10(44 - R) = $ _____

If $R \geq 44$, $S = 800$.

ANSWER SHEET
Practice Test 5

1. Ⓐ Ⓑ Ⓒ Ⓓ Ⓔ
2. Ⓐ Ⓑ Ⓒ Ⓓ Ⓔ
3. Ⓐ Ⓑ Ⓒ Ⓓ Ⓔ
4. Ⓐ Ⓑ Ⓒ Ⓓ Ⓔ
5. Ⓐ Ⓑ Ⓒ Ⓓ Ⓔ
6. Ⓐ Ⓑ Ⓒ Ⓓ Ⓔ
7. Ⓐ Ⓑ Ⓒ Ⓓ Ⓔ
8. Ⓐ Ⓑ Ⓒ Ⓓ Ⓔ
9. Ⓐ Ⓑ Ⓒ Ⓓ Ⓔ
10. Ⓐ Ⓑ Ⓒ Ⓓ Ⓔ
11. Ⓐ Ⓑ Ⓒ Ⓓ Ⓔ
12. Ⓐ Ⓑ Ⓒ Ⓓ Ⓔ
13. Ⓐ Ⓑ Ⓒ Ⓓ Ⓔ

14. Ⓐ Ⓑ Ⓒ Ⓓ Ⓔ
15. Ⓐ Ⓑ Ⓒ Ⓓ Ⓔ
16. Ⓐ Ⓑ Ⓒ Ⓓ Ⓔ
17. Ⓐ Ⓑ Ⓒ Ⓓ Ⓔ
18. Ⓐ Ⓑ Ⓒ Ⓓ Ⓔ
19. Ⓐ Ⓑ Ⓒ Ⓓ Ⓔ
20. Ⓐ Ⓑ Ⓒ Ⓓ Ⓔ
21. Ⓐ Ⓑ Ⓒ Ⓓ Ⓔ
22. Ⓐ Ⓑ Ⓒ Ⓓ Ⓔ
23. Ⓐ Ⓑ Ⓒ Ⓓ Ⓔ
24. Ⓐ Ⓑ Ⓒ Ⓓ Ⓔ
25. Ⓐ Ⓑ Ⓒ Ⓓ Ⓔ
26. Ⓐ Ⓑ Ⓒ Ⓓ Ⓔ

27. Ⓐ Ⓑ Ⓒ Ⓓ Ⓔ
28. Ⓐ Ⓑ Ⓒ Ⓓ Ⓔ
29. Ⓐ Ⓑ Ⓒ Ⓓ Ⓔ
30. Ⓐ Ⓑ Ⓒ Ⓓ Ⓔ
31. Ⓐ Ⓑ Ⓒ Ⓓ Ⓔ
32. Ⓐ Ⓑ Ⓒ Ⓓ Ⓔ
33. Ⓐ Ⓑ Ⓒ Ⓓ Ⓔ
34. Ⓐ Ⓑ Ⓒ Ⓓ Ⓔ
35. Ⓐ Ⓑ Ⓒ Ⓓ Ⓔ
36. Ⓐ Ⓑ Ⓒ Ⓓ Ⓔ
37. Ⓐ Ⓑ Ⓒ Ⓓ Ⓔ
38. Ⓐ Ⓑ Ⓒ Ⓓ Ⓔ
39. Ⓐ Ⓑ Ⓒ Ⓓ Ⓔ

40. Ⓐ Ⓑ Ⓒ Ⓓ Ⓔ
41. Ⓐ Ⓑ Ⓒ Ⓓ Ⓔ
42. Ⓐ Ⓑ Ⓒ Ⓓ Ⓔ
43. Ⓐ Ⓑ Ⓒ Ⓓ Ⓔ
44. Ⓐ Ⓑ Ⓒ Ⓓ Ⓔ
45. Ⓐ Ⓑ Ⓒ Ⓓ Ⓔ
46. Ⓐ Ⓑ Ⓒ Ⓓ Ⓔ
47. Ⓐ Ⓑ Ⓒ Ⓓ Ⓔ
48. Ⓐ Ⓑ Ⓒ Ⓓ Ⓔ
49. Ⓐ Ⓑ Ⓒ Ⓓ Ⓔ
50. Ⓐ Ⓑ Ⓒ Ⓓ Ⓔ

Practice Test 5

T ear out the preceding answer sheet. Decide which is the best choice by rounding your answer when appropriate. Blacken the corresponding space on the answer sheet. When finished, check your answers with those at the end of the test. For questions that you got wrong, note the chapters containing the material that you must review. Also, if you do not fully understand how you arrived at some of the correct answers, you should review the appropriate chapters. Finally, fill out the Self-Evaluation Chart on page 319 in order to pinpoint the topics that gave you the most difficulty.

50 questions: 1 hour

Directions: Decide which answer choice is best. If the exact numerical value is not one of the answer choices, select the closest approximation. Fill in the circle on the answer sheet that corresponds to your choice.

Notes:
(1) You will need to use a scientific or graphing calculator to answer some of the questions.
(2) You will have to decide whether to put your calculator in degree or radian mode for some problems.
(3) All figures that accompany problems are plane figures unless otherwise stated. Figures are drawn as accurately as possible to provide useful information for solving the problem, except when it is stated in a particular problem that the figure is not drawn to scale.
(4) Unless otherwise indicated, the domain of a function is the set of all real numbers for which the functional value is also a real number.

Reference Information. The following formulas are provided for your information.

Volume of a right circular cone with radius r and height h: $V = \dfrac{1}{3}\pi r^2 h$

Volume of a sphere of radius r: $V = \dfrac{4}{3}\pi r^3$

Surface area of a sphere of radius r: $S = 4\pi r^2$

Volume of a pyramid of base area B and height h: $V = \dfrac{1}{3}Bh$

PRACTICE TEST 5

1. $x^{2/3} + x^{4/3} =$

   (A) $x^{2/3}$
   (B) $x^{8/9}$
   (C) x
   (D) x^2
   (E) $x^{2/3}(x^{2/3} + 1)$

2. In three dimensions, what is the set of all points for which $x = 0$?

   (A) the origin
   (B) a line parallel to the x-axis
   (C) the yz-plane
   (D) a plane that contains the x-axis
   (E) the x-axis

3. Greg deposits $1,000 into a savings account that earns 2% interest compounded annually. If Greg makes no additional deposits or withdrawals, approximately how many years will it take for the amount in the account to double?

   (A) 10
   (B) 15
   (C) 28
   (D) 36
   (E) 42

4. If $f(x) = \sqrt[3]{x}$ and $g(x) = x^3 + 8$, find $(f \circ g)(3)$.

   (A) 3.3
   (B) 5
   (C) 11
   (D) 35
   (E) 50.5

5. $x > \sin x$ for

   (A) all $x > 0$
   (B) all $x < 0$
   (C) all x for which $x \neq 0$
   (D) all x
   (E) all x for which $-\dfrac{\pi}{2} < x < 0$

6. The sum of the zeros of $f(x) = 3x^2 - 5$ is

   (A) 3.3
   (B) 1.8
   (C) 1.7
   (D) 1.3
   (E) 0

7. The intersection of a plane with a right circular cylinder could be which of the following?

   I. A circle
  II. Parallel lines
 III. Intersecting lines

(A) I only
(B) II only
(C) III only
(D) I and II only
(E) I, II, and III

8. Two dice are tossed. What is the probability that the sum is 5?

(A) $\dfrac{1}{11}$

(B) $\dfrac{1}{9}$

(C) $\dfrac{1}{6}$

(D) $\dfrac{1}{4}$

(E) $\dfrac{1}{2}$

9. The graph of $f(x) = \dfrac{10}{x^2 - 10x + 25}$ has a vertical asymptote at $x =$

(A) 0 only
(B) 5 only
(C) 10 only
(D) 0 and 5 only
(E) 0, 5, and 10

10. When $P(x) = 2x^3 - 6x^2 + Kx$ is divided by $x + 2$, the remainder is –10. Then $K =$

(A) –30
(B) –15
(C) –6
(D) –1
(E) 6

11. Of the following lists of numbers, which has the largest standard deviation?

   (A) 2, 7, 15
   (B) 3, 7, 14
   (C) 5, 7, 12
   (D) 10, 11, 12
   (E) 11, 11, 11

USE THIS SPACE FOR SCRATCH WORK

12. If $f(x)$ is a linear function and $f(2) = 1$ and $f(4) = -2$, then $f(x) =$

   (A) $-\dfrac{3}{2}x + 4$

   (B) $\dfrac{3}{2}x - 2$

   (C) $-\dfrac{3}{2}x + 2$

   (D) $\dfrac{3}{2}x - 4$

   (E) $-\dfrac{2}{3}x + \dfrac{7}{3}$

13. The length of the radius of a circle is one-half the length of an arc of the circle. What is the radian measure of the central angle that intercepts that arc?

   (A) $60°$
   (B) $120°$
   (C) 1^R
   (D) 2^R
   (E) π^R

14. If $f(x) = 2^x + 1$, then $f^{-1}(7) =$

   (A) 2.4
   (B) 2.6
   (C) 2.8
   (D) 3
   (E) 3.6

15. Find all values of x that satisfy the determinant equation $\begin{vmatrix} 2x & 1 \\ x & x \end{vmatrix} = 3$.

   (A) -1
   (B) -1 or 1.5
   (C) 1.5
   (D) -1.5
   (E) -1.5 or 1

16. The 71st term of $30, 27, 24, 21, \ldots$, is

(A) 5,325
(B) 240
(C) 180
(D) −180
(E) −183

17. If $0 < x < \dfrac{\pi}{2}$ and $\tan 5x = 3$, to the nearest tenth,

what is the value of $\tan x$?

(A) 0.5
(B) 0.4
(C) 0.3
(D) 0.2
(E) 0.1

18. If $4.05^p = 5.25^q$, what is the value of $\dfrac{p}{q}$?

(A) −0.11
(B) 0.11
(C) 1.19
(D) 1.30
(E) 1.67

19. A cylinder has a base radius of 2 and a height of 9. To the nearest whole number, by how much does the lateral area exceed the sum of the areas of the two bases?

(A) 101
(B) 96
(C) 88
(D) 81
(E) 75

20. If $\cos 67° = \tan x°$, then $x =$

(A) 0.4
(B) 6.8
(C) 7.8
(D) 21
(E) 29.3

21. $P(x) = x^3 + 18x - 30$ has a zero in the interval

(A) (0,0.5)
(B) (0.5,1)
(C) (1,1.5)
(D) (1.5,2)
(E) (2,2.5)

22. The lengths of the sides of a triangle are 23, 32, and 37. To the nearest degree, what is the value of the largest angle?

(A) 71°
(B) 83°
(C) 122°
(D) 128°
(E) 142°

23. If $f(x) = \dfrac{3}{x-2}$ and $g(x) = \sqrt{x+1}$, find the domain of $f \circ g$.

(A) $x \geq -1$
(B) $x \neq 2$
(C) $x \geq -1, x \neq 2$
(D) $x \geq -1, x \neq 3$
(E) $x \leq -1$

24. The overall average grade of a math course is based on homework (10%), quizzes (40%), and tests (50%). Ted has a 90 average for homework, 81 for quizzes, and 85 for tests. What is his overall average?

(A) 83
(B) 84
(C) 85
(D) 86
(E) 87

25. If $y^{\frac{1}{2}} = 3.216$, then $(10y)^{\frac{1}{2}} =$

(A) 321.6
(B) 32.16
(C) 10.17
(D) 5.67
(E) 4.23

26. What is the domain of the function
$f(x) = \log \sqrt{2x^2 - 15}$?

(A) $-7.5 < x < 7.5$
(B) $x < -7.5$ or $x > 7.5$
(C) $x < -2.7$ or $x > 2.7$
(D) $x < -3.2$ or $x > 3.2$
(E) $x < 1.9$ or $x > 1.9$

USE THIS SPACE FOR SCRATCH WORK

27. A magazine has 1,200,000 subscribers, of whom 400,000 are women and 800,000 are men. Twenty percent of the women and 60 percent of the men read the advertisements in the magazine. What is the probability that a randomly selected subscriber reads the advertisements?

(A) 0.30
(B) 0.36
(C) 0.40
(D) 0.47
(E) 0.52

28. Let S be the sum of the first n terms of the arithmetic sequence $3, 7, 11, \ldots$, and let T be the sum of the first n terms of the arithmetic sequence $8, 10, 12, \ldots$. For $n > 1$, $S = T$ for

(A) no value of n
(B) one value of n
(C) two values of n
(D) three values of n
(E) four values of n

29. On the interval $\left[-\dfrac{\pi}{4}, \dfrac{\pi}{4}\right]$, the function

$f(x) = \sqrt{1 + \sin^2 x}$ has a maximum value of

(A) 0.78
(B) 1
(C) 1.1
(D) 1.2
(E) 1.4

30. A point has rectangular coordinates (3,4). The polar coordinates are (5,θ). What is the value of θ?

(A) 30°
(B) 37°
(C) 51°
(D) 53°
(E) 60°

31. If $f(x) = x^2 - 4$, for what real number values of x will $f(f(x)) = 0$?

(A) 2.4
(B) ±2.4
(C) 2 or 6
(D) ±1.4 or ±2.4
(E) no values

32. If $f(x) = x \log x$ and $g(x) = 10^x$, then $g(f(2)) =$

(A) 24
(B) 17
(C) 4
(D) 2
(E) 0.6

33. If $f(x) = x^{\sqrt{x}}$, then $f\left(\sqrt{2}\right) =$

(A) 1.4
(B) 1.5
(C) 1.6
(D) 2.0
(E) 2.7

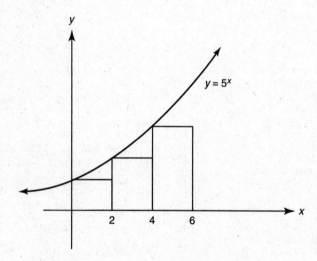

34. The figure above shows the graph of 5^x. What is the sum of the areas of the rectangles?

(A) 32,550
(B) 16,225
(C) 2,604
(D) 1,302
(E) 651

35. (p,q) is called a *lattice point* if p and q are both integers. How many lattice points lie in the area strictly between the two curves $x^2 + y^2 = 9$ and $x^2 + y^2 - 6x + 5 = 0$?

(A) 0
(B) 1
(C) 2
(D) 3
(E) 4

36. If $9^x = \sqrt{3}$ and $2^{x+y} = 32$, then $y =$

(A) $\dfrac{1}{2}$

(B) $\dfrac{3}{4}$

(C) $\dfrac{5}{2}$

(D) $\dfrac{11}{2}$

(E) $\dfrac{19}{4}$

37. For all real numbers x, $f(2x) = x^2 - x + 3$. An expression for $f(x)$ in terms of x is

(A) $2x^2 - 2x + 3$

(B) $4x^2 - 2x + 3$

(C) $\dfrac{x^2}{4} - \dfrac{x}{2} + 3$

(D) $\dfrac{x^2}{2} - \dfrac{x}{2} + 3$

(E) $x^2 - x + 3$

38. For what value(s) of k is $x^2 - kx + k$ divisible by $x - k$?

(A) only 0

(B) only 0 or $-\dfrac{1}{2}$

(C) only 1
(D) any value of k
(E) no value of k

39. If the graphs of $x^2 = 4(y + 9)$ and $x + ky = 6$ intersect on the x-axis, then $k =$

(A) 0
(B) 6
(C) –6
(D) no real number
(E) any real number

40. The length of the major axis of the ellipse $\dfrac{(x-3)^2}{16} + \dfrac{(y+2)^2}{25} = 1$ is

(A) 3
(B) 4
(C) 5
(D) 6
(E) 10

41. If $f_n = \begin{cases} \dfrac{f_{n-1}}{2} & \text{when } f_{n-1} \text{ is an even number} \\[2mm] 3 \cdot f_{n-1} + 1 & \text{when } f_{n-1} \text{ is an odd number} \end{cases}$

and $f_1 = 3$, then $f_5 =$

(A) 1
(B) 2
(C) 4
(D) 8
(E) 16

42. How many different rearrangements of the letters in the word CONTEST start with the two vowels?

(A) 120
(B) 60
(C) 10
(D) 5
(E) none of the above

43. Which of the following translations of the graph of $y = x^2$ would result in the graph of $y = x^2 - 6x + k$, where k is a constant greater than 10?

(A) left 6 units and up k units
(B) left 3 units and up $k + 9$ units
(C) right 3 units and up $k + 9$ units
(D) left 3 units and up $k - 9$ units
(E) right 3 units and up $k - 9$ units

44. How many positive integers are there in the solution set of $\dfrac{x}{x-2} > 5$?

(A) 0
(B) 2
(C) 4
(D) 5
(E) an infinite number

45. Which of the following is equivalent to $3x^2 - x < 2$?

(A) $-\dfrac{3}{2} < x < 1$

(B) $-1 < x < \dfrac{2}{3}$

(C) $-\dfrac{2}{3} < x < 1$

(D) $-1 < x < \dfrac{3}{2}$

(E) $x < -\dfrac{2}{3}$ or $x > 1$

46. If $x_0 = 3$ and $x_{n+1} = x_n\sqrt{x_n + 1}$, then $x_3 =$

(A) 15.9
(B) 31.7
(C) 44.9
(D) 65.2
(E) 173.9

47. When the smaller root of the equation $3x^2 + 4x - 1 = 0$ is subtracted from the larger root, the result is

(A) −1.3
(B) 0.7
(C) 1.3
(D) 1.8
(E) 2.0

48. A committee of 4 is to be selected from a group of 7 women and 4 men. How many different committees of 2 men and 2 women can be formed?

(A) 22
(B) 24
(C) 126
(D) 128
(E) 789

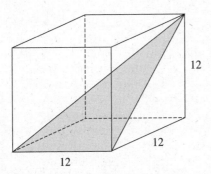

49. Find the area of the shaded region formed within the cube of edge length 12 shown in the figure above.

(A) 50.91
(B) 95.22
(C) 101.82
(D) 124.87
(E) 203.65

50. In radians, $\sin^{-1}(\cos 100°) =$

(A) −1.4
(B) −0.2
(C) 0.2
(D) 1.0
(E) 1.4

STOP

If there is still time remaining, you may review your answers.

1. **E**	14. **B**	27. **D**	40. **E**
2. **C**	15. **B**	28. **B**	41. **D**
3. **D**	16. **D**	29. **D**	42. **A**
4. **A**	17. **C**	30. **D**	43. **E**
5. **A**	18. **C**	31. **D**	44. **A**
6. **E**	19. **C**	32. **C**	45. **C**
7. **D**	20. **D**	33. **B**	46. **D**
8. **B**	21. **C**	34. **D**	47. **D**
9. **B**	22. **B**	35. **D**	48. **C**
10. **B**	23. **D**	36. **E**	49. **C**
11. **A**	24. **B**	37. **C**	50. **B**
12. **A**	25. **C**	38. **A**	
13. **D**	26. **C**	39. **E**	

ANSWERS EXPLAINED

The following explanations are keyed to the review portions of this book. The number in brackets after each explanation indicates the appropriate chapter in the Review of Major Topics (Part 2).

An asterisk appears next to those solutions for which a graphing calculator is necessary or helpful.

1. **(E)** Factor out $x^{2/3}$, the greatest common factor of $x^{2/3} + x^{4/3}$, to get:

$$x^{2/3} + x^{4/3} = x^{2/3}(1 + x^{2/3}) \ [8]$$

2. **(C)** When $x = 0$, y and z can be any value. Therefore, any point in the yz-plane is a possible member of the set. [15]

3. **(D)** A deposit of $1,000 that earns a 2% annual interest rate can be modeled by the growth equation $A = P(1.02)^t$, where P is the initial amount invested and A is the value of the account after t years. In this case, $P = 1,000$, $A = 2,000$, and we need to find t: $2,000 = 1,000(1.02)^t$ or $2 = (1.02)^t$. Take the log of both sides to get $\log 2 = t \log 1.02$ or $t = \dfrac{\log 2}{\log 1.02} = 35.0027$. It takes just under a day into the 36th year before the amount is doubled. [8]

* 4. **(A)** Enter f in Y_1 and g in Y_2, return to the home screen, and enter $Y_1(Y_2(3))$ to get the correct answer choice. [1]

* 5. **(A)** Plot the graph of $y = x - \sin x$ and observe that the graph lies above the x-axis for all $x > 0$. [7]

* 6. **(E)** Plot the graph of $y = 3x^2 - 5$ in the standard window. The symmetry about the y-axis indicates that the zeros are opposites and therefore sum to zero.

    An alternative solution is to use the fact that the zeros of a quadratic function sum to

    $-\dfrac{b}{a} = 0$. [3]

7. **(D)** If the plane is perpendicular to the axis of the cylinder, the intersection will be a circle, so I is possible. If the plane is parallel to the axis of the cylinder, the intersection will be parallel lines, so II is possible. It is impossible for the intersection of a plane and a cylinder to form intersecting lines because there are no intersecting lines on a cylinder. [15]

8. **(B)** Call the outcome of a roll (A,B). The sum of A and B is 5 for $(1,4)$, $(4,1)$, $(2,3)$, and $(3,2)$. There are $36 = (6)(6)$ possible outcomes. Therefore, the probability of a roll of 5 is $\dfrac{4}{36} = \dfrac{1}{9}$. [22]

* 9. **(B)** Plot f in a standard window and use TRACE to observe that y increases without bound on either side of $x = 5$.

    An alternative solution can be found by observing that the denominator of f is $(x - 5)^2$, which equals zero only when $x = 5$. [9]

* 10. **(B)** By the Remainder Theorem, substituting -2 for x yields the remainder:

$$-16 - 24 - 2K = -10$$
$$-40 - 2K = -10$$
$$-2K = 30$$
$$K = -15 \ [4]$$

11. **(A)** The standard deviation of a data set is a measure of the spread in the data. Of the five data sets presented, Choice A exhibits the greatest spread and therefore has the greatest standard deviation. [21]

12. **(A)** The slope of $f(x) = \dfrac{-2-1}{4-2} = \dfrac{-3}{2}$. Using the point-slope form, $f(x) - 1 =$

$\dfrac{-3}{2}(x - 2)$. Therefore, $f(x) = \dfrac{-3}{2}x + 4$. [2]

13. **(D)** $s = r\theta$. $2r = r\theta$. $\theta = 2$. [7]

* 14. **(B)** Plot the graphs of $y = 2^x + 1$ in the standard window. Since $f^{-1}(7)$ is the value of x such that $f(x) = 7$, also plot the graph of $y = 7$. The correct answer choice is the point where these two graphs intersect.

An alternative solution is to solve the equation $2^x + 1 = 7$: $x = \log_2 6 = \dfrac{\log 6}{\log 2} \approx 2.6$. [1]

15. **(B)** Evaluate the 2 by 2 determinant to get $2x^2 - x = 3$. Then factor $2x^2 - x - 3$ into

$(2x - 3)(x + 1)$ and set each factor equal to zero to get $x = 1.5$ or $x = -1$. [18]

* 16. **(D)** Use LIST/seq to construct the sequence as seq$(30 - 3x, x, 0, 70)$ and store this sequence in a list (e.g., L_1). On the home screen, enter $L_1(71)$ for the 71st term, -180.

An alternative solution is to use the formula for the nth term of an arithmetic sequence:

$$t_{71} = 30 - (70)(3) = -180 \ [19]$$

* 17. **(C)** Put your calculator in radian mode: $5x = \tan^{-1} 3 \approx 1.249$. $x \approx 0.2498$. Therefore, $\tan x = 0.255139 \approx 0.3$. [7]

* 18. **(C)** Take $\log_{4.05}$ of both sides of the equation, getting $p = \log_{4.05} 5.25^q$, or $p = q \log_{4.05} 5.25$. Dividing both sides by q and changing to base 10 yields:

$$\frac{p}{q} = \frac{\log 5.25}{\log 4.05} \approx 1.19 \ [8]$$

* 19. **(C)**

$$\text{Area of one base} = \pi r^2 = 4\pi$$
$$\text{Lateral area} = 2\pi r h = 36\pi$$
$$\text{Lateral area} - \text{two bases} = 36\pi - 8\pi = 28\pi \approx 87.96 \approx 88 \ [15]$$

* 20. **(D)** Put your calculator in degree mode and evaluate $\tan^{-1}(\cos 67°)$. [7]

* 21. **(C)** Plot the graph of $y = x^3 + 18x - 30$ in a $[0,3]$ by $[-5,5]$ window and use CALC/zero to locate a zero at $x \approx 1.48$. [4]

✱ 22. **(B)** The angle opposite the 37 side (call it $\angle A$) is the largest angle. By the Law of Cosines, $37^2 = 23^2 + 32^2 - 2(23)(32) \cos A$.

$$\cos A = \frac{37^2 - 23^2 - 32^2}{-2(23)(32)} \approx \frac{-184}{-1,472} = 0.125$$

Therefore, $A = \cos^{-1}(0.125) \approx 83°$. [6]

23. **(D)** The domain of g is $x \geq -1$ because $x + 1 \geq 0$. The domain of $f \circ g$ consists of all x such that $g(x)$ is in the domain of f. Since $g(3) = 2$ and 2 is not in the domain of f, 3 must be excluded from the domain of $f \circ g$. [1]

✱ 24. **(B)** Calculate the weighted average: $(0.1)(90) + (0.4)(81) + (0.5)(85) = 83.9$, which rounds to 84. [21]

✱ 25. **(C)** Since $y^{\frac{1}{2}} = 3.216$, $y \approx 10.34$, and $10y \approx 103.4$, so $(10y)^{\frac{1}{2}} \approx 10.17$. [8]

✱ 26. **(C)** Plot the graph of $y = \log \sqrt{2x^2 - 15}$ in the standard window and zoom in once. Use the TRACE to determine that there are no y-values on the graph between the approximate x-values of -2.7 and 2.7.

An alternative solution is to use the fact that the domain of the square root function is $x \geq 0$ and its range is $y \geq 0$. However, since the domain of the log function is $x > 0$, the domain of the function f must satisfy $2x^2 - 15 > 0$, or $x^2 > 7.5$. The approximate solution to this inequality is the correct answer, Choice C. [8]

✱ 27. **(D)** Twenty percent of 400,000 women is 80,000, and 60% of 800,000 men is 480,000. Altogether 560,000 subscribers, or about 47%, read ads. [22]

28. **(B)** Equate S and T using the formula for the sum of the first n terms of an arithmetic series: $\frac{n}{2}[6 + (n-1)4] = \frac{n}{2}[16 + (n-1)2]$. This equation reduces to $n^2 - 6n = 0$, for which 6 is the only positive solution. [19]

✱ 29. **(D)** Plot the graph of $y = \sqrt{(1 + \sin(x)^2)}$ in a $\left[-\dfrac{\pi}{4}, \dfrac{\pi}{4}\right]$ by $[-2,2]$ window and observe that the maximum value of y occurs at the endpoints. Return to the home screen, and enter $Y_1(\pi/4)$ to get the correct answer choice.

An alternative solution uses the fact that on the interval $\left[-\dfrac{\pi}{4}, \dfrac{\pi}{4}\right]$, the maximum value

of $|\sin x| = \dfrac{\sqrt{2}}{2}$. Therefore, the maximum value of $f(x)$ is $\sqrt{1 + \dfrac{1}{2}} = \sqrt{\dfrac{3}{2}} \approx 1.2$. [7]

* 30.　　**(D)** $\sin\theta = \dfrac{4}{5}$

Therefore, $\theta = \sin^{-1}\left(\dfrac{4}{5}\right) \approx 53°$. [14]

* 31.　　**(D)** Plot the graph of $f(f(x))$ in the standard window by entering $Y_1 = x^2 - 4$, $Y_2 = Y_1(x)$, and de-selecting Y_1. The graph has 4 zeros located symmetrically about the y-axis, making Choice D the only possible solution.

An alternative solution is to evaluate $f(f(x)) = (x^2 - 4)^2 - 4 = 0$ and solve by setting $x^2 - 4 = \pm 2$, or $x^2 = 6$ or 2, so $x = \pm\sqrt{6}$ or $\pm\sqrt{2}$. Again, Choice D is the only answer choice with 4 solutions. [1]

* 32.　　**(C)** Enter f into Y_1 and g into Y_2. Evaluate $Y_2(Y_1(2))$ for the correct answer choice.

An alternative solution is to use the properties of logs to evaluate $f(2) = 2 \log 2 = \log 4$ and $g(\log 4) = 10^{\log 4} = 4$ without a calculator. [8]

* 33.　　**(B)** $f\left(\sqrt{2}\right) = \left(\sqrt{2}\right)^{\sqrt{2}} \approx 1.5$. [1]

* 34.　　**(D)** The width of each rectangle is 2 and the heights are $5^0 = 1$, $5^2 = 25$, and $5^4 = 625$. Therefore, the total area is $2(1 + 25 + 625) = 1{,}302$. [8]

* 35.　　**(D)** Plot the graphs of $y = \pm\sqrt{9 - x^2}$ and $y = \pm\sqrt{-x^2 + 6x - 5}$ in the standard window, but with FORMAT set to GridOn. The "grid" consists exactly of the lattice points. ZOOM/ZBox around the area enclosed by the two graphs, and count the number of lattice points in that area to be 3. The points $(1,0)$ and $(3,0)$ appear close to the boundary, but a mental check finds that $(1,0)$ is on the boundary of the second curve, while $(3,0)$ is on the boundary of the first. [15]

* 36.　　**(E)** $9^x = (3^2)^x = 3^{2x}$ and $\sqrt{3} = 3^{1/2}$. Therefore, $2x = \dfrac{1}{2}$ and $x = \dfrac{1}{4}$. Since $32 = 2^5$, then $x + y = 5$ and $y = \dfrac{19}{4}$. [8]

　37.　　**(C)** $f(x) = f\left(2 \cdot \dfrac{x}{2}\right) = \left(\dfrac{x}{2}\right)^2 - \dfrac{x}{2} + 3 = \dfrac{x^2}{4} - \dfrac{x}{2} + 3$. [1]

　38.　　**(A)** Using the factor theorem, substitute k for x and set the result equal to zero. Then $k^2 - k^2 + k = 0$, and $k = 0$. [4]

　39.　　**(E)** If the graphs intersect on the x-axis, the value of y must be zero. Since the value of y is zero, it does not matter what k is. [4]

40. **(E)** This ellipse has its center at $(3,-2)$. Since $25 > 16$, the major axis is vertical, and its length is $2\sqrt{25} = 10$. [13]

41. **(D)**

n	1	2	3	4	5
f_n	3	10	5	16	8

[19]

* 42. **(A)** There are 5 consonants, CNTST, but the two Ts are indistinguishable, so there are $\dfrac{5!}{2} = 60$ ways of arranging these. There are two ways of arranging the 2 vowels in the front. Therefore, there are $2 \cdot 60 = 120$ distinguishable arrangements. [16]

43. **(E)** Complete the square on $x^2 - 6x$ by adding 9. Then:

$$x^2 - 6x + k = x^2 - 6x + 9 + k - 9 = (x - 3)^2 + (k - 9)$$

This expression represents the translation of x^2 by 3 units right and $k - 9$ units up. [12]

* 44. **(A)** Plot the graph of $y = \dfrac{x}{x - 2} - 5$ in the standard window and observe the vertical asymptote at $x = 2$. $y > \dfrac{x}{x - 2} - 5$ where this graph lies above the x-axis, and there are no integer values of x in this interval.

An alternative solution is to solve the equation $\dfrac{x}{x - 2} = 5 \ (x \neq 2)$: $x = 5x - 10$, or $x = \dfrac{5}{2}$. If you test values in the intervals $(-\infty, 2)$, $\left(2, \dfrac{5}{2}\right)$, and $\left(\dfrac{5}{2}, \infty\right)$ you find that only the middle interval satisfies the inequality, but it contains no integers. [5]

45. **(C)** Subtract 2 from both sides of the inequality to get $3x^2 - x - 2 < 0$, and solve the corresponding equation by factoring the left side: $(3x + 2)(x - 1) = 0$. The solutions to this equation are $x = -\dfrac{2}{3}$ and $x = 1$. Select test points in each of the regions: $x < -\dfrac{2}{3}$, $-\dfrac{2}{3} < x < 1$, and $x > 1$ to determine that $3x^2 - x - 2 < 0$. Therefore, $3x^2 - x < 2$ when $-\dfrac{2}{3} < x < 1$. [1]

* 46. **(D)** Enter 3 into your calculator. Then enter Ans $\sqrt{\text{Ans}+1}$ three times to accomplish three iterations that result in x_3 and the correct answer choice. [19]

* 47. **(D)** Substitute the values of a, b, and c into the Quadratic Formula to find the algebraic solutions; then subtract the smaller from the larger to find the decimal approximation to the answer:

$$x = \frac{-4 \pm \sqrt{16 + 12}}{6} = \frac{-2 \pm \sqrt{7}}{3}; \quad \frac{-2 + \sqrt{7}}{3} - \frac{-2 - \sqrt{7}}{3} = \frac{2\sqrt{7}}{3} \approx 1.8 \quad [3]$$

* **48.**   **(C)** There are $\begin{pmatrix} 4 \\ 2 \end{pmatrix} = \dfrac{4 \cdot 3}{2} = 6$ ways of selecting 2 men of 4 and $\begin{pmatrix} 7 \\ 2 \end{pmatrix} = \dfrac{7 \cdot 6}{2} = 21$

ways of selecting 2 women of 7. Therefore, there are $\begin{pmatrix} 4 \\ 2 \end{pmatrix}\begin{pmatrix} 7 \\ 2 \end{pmatrix} = 6 \cdot 21 = 126$ ways of

selecting a committee that consists of 2 men and 2 women. [16]

49.   **(C)** The region is a triangle, formed by the lower front edge, the diagonal of the right face, and the diagonal from the lower left corner to the upper right corner. The angle of the triangle at the lower right front corner is a right angle. Therefore, one base of the triangle has length 12 and the height that corresponds to this base is $12\sqrt{2}$. Therefore, the area of the triangle is $\dfrac{1}{2}(12)\left(12\sqrt{2}\right) \approx 101.82.$ [15]

* **50.**   **(B)** With your calculator in degree mode, evaluate $\sin^{-1}(\cos 100°) = -10°$ directly. To change to radians, multiply $-10°$ by $\dfrac{\pi}{180°}$ to get $-0.17 \approx -0.2$ radians. [7]

Self-Evaluation Chart for Practice Test 5

Subject Area	Questions and Review Section								Right	Number Wrong	Omitted
Algebra and Functions (24 questions)	1 8	3 8	4 1	6 3	9 9				___	___	___
	10 4	12 2	14 1	18 8	21 4	23 1			___	___	___
	25 8	26 8	31 1	32 8	33 1	34 8	36 8		___	___	___
	37 1	38 4	39 4	44 5	45 1	47 3			___	___	___
Trigonometry (7 questions)	5 7	13 7	17 7	20 7	22 6	29 7	50 7		___	___	___
Coordinate and Three-Dimensional Geometry (8 questions)	2 15	7 15	19 15	30 14	35 15	40 13	43 12	49 15	___	___	___
Number and Operations (7 questions)	15 18	16 19	28 19	41 19	42 16	46 19	48 16		___	___	___
Data Analysis, Statistics, and Probability (4 questions)	8 22	11 21	24 21	27 22					___	___	___
TOTALS									___	___	___

Evaluate Your Performance on Practice Test 5	
Rating	**Number Right**
Excellent	41–50
Very good	33–40
Above average	25–32
Average	15–24
Below average	Below 15

CALCULATING YOUR SCORE

Raw score R = number right $- \dfrac{1}{4}$(number wrong), rounded = _____

Approximate scaled score $S = 800 - 10(44 - R) =$ _____

If $R \geq 44$, $S = 800$.

1. Ⓐ Ⓑ Ⓒ Ⓓ Ⓔ	14. Ⓐ Ⓑ Ⓒ Ⓓ Ⓔ	27. Ⓐ Ⓑ Ⓒ Ⓓ Ⓔ	40. Ⓐ Ⓑ Ⓒ Ⓓ Ⓔ
2. Ⓐ Ⓑ Ⓒ Ⓓ Ⓔ	15. Ⓐ Ⓑ Ⓒ Ⓓ Ⓔ	28. Ⓐ Ⓑ Ⓒ Ⓓ Ⓔ	41. Ⓐ Ⓑ Ⓒ Ⓓ Ⓔ
3. Ⓐ Ⓑ Ⓒ Ⓓ Ⓔ	16. Ⓐ Ⓑ Ⓒ Ⓓ Ⓔ	29. Ⓐ Ⓑ Ⓒ Ⓓ Ⓔ	42. Ⓐ Ⓑ Ⓒ Ⓓ Ⓔ
4. Ⓐ Ⓑ Ⓒ Ⓓ Ⓔ	17. Ⓐ Ⓑ Ⓒ Ⓓ Ⓔ	30. Ⓐ Ⓑ Ⓒ Ⓓ Ⓔ	43. Ⓐ Ⓑ Ⓒ Ⓓ Ⓔ
5. Ⓐ Ⓑ Ⓒ Ⓓ Ⓔ	18. Ⓐ Ⓑ Ⓒ Ⓓ Ⓔ	31. Ⓐ Ⓑ Ⓒ Ⓓ Ⓔ	44. Ⓐ Ⓑ Ⓒ Ⓓ Ⓔ
6. Ⓐ Ⓑ Ⓒ Ⓓ Ⓔ	19. Ⓐ Ⓑ Ⓒ Ⓓ Ⓔ	32. Ⓐ Ⓑ Ⓒ Ⓓ Ⓔ	45. Ⓐ Ⓑ Ⓒ Ⓓ Ⓔ
7. Ⓐ Ⓑ Ⓒ Ⓓ Ⓔ	20. Ⓐ Ⓑ Ⓒ Ⓓ Ⓔ	33. Ⓐ Ⓑ Ⓒ Ⓓ Ⓔ	46. Ⓐ Ⓑ Ⓒ Ⓓ Ⓔ
8. Ⓐ Ⓑ Ⓒ Ⓓ Ⓔ	21. Ⓐ Ⓑ Ⓒ Ⓓ Ⓔ	34. Ⓐ Ⓑ Ⓒ Ⓓ Ⓔ	47. Ⓐ Ⓑ Ⓒ Ⓓ Ⓔ
9. Ⓐ Ⓑ Ⓒ Ⓓ Ⓔ	22. Ⓐ Ⓑ Ⓒ Ⓓ Ⓔ	35. Ⓐ Ⓑ Ⓒ Ⓓ Ⓔ	48. Ⓐ Ⓑ Ⓒ Ⓓ Ⓔ
10. Ⓐ Ⓑ Ⓒ Ⓓ Ⓔ	23. Ⓐ Ⓑ Ⓒ Ⓓ Ⓔ	36. Ⓐ Ⓑ Ⓒ Ⓓ Ⓔ	49. Ⓐ Ⓑ Ⓒ Ⓓ Ⓔ
11. Ⓐ Ⓑ Ⓒ Ⓓ Ⓔ	24. Ⓐ Ⓑ Ⓒ Ⓓ Ⓔ	37. Ⓐ Ⓑ Ⓒ Ⓓ Ⓔ	50. Ⓐ Ⓑ Ⓒ Ⓓ Ⓔ
12. Ⓐ Ⓑ Ⓒ Ⓓ Ⓔ	25. Ⓐ Ⓑ Ⓒ Ⓓ Ⓔ	38. Ⓐ Ⓑ Ⓒ Ⓓ Ⓔ	
13. Ⓐ Ⓑ Ⓒ Ⓓ Ⓔ	26. Ⓐ Ⓑ Ⓒ Ⓓ Ⓔ	39. Ⓐ Ⓑ Ⓒ Ⓓ Ⓔ	

PRACTICE TEST 6

Practice Test 6

Tear out the preceding answer sheet. Decide which is the best choice by rounding your answer when appropriate. Blacken the corresponding space on the answer sheet. When finished, check your answers with those at the end of the test. For questions that you got wrong, note the chapters containing the material that you must review. Also, if you do not fully understand how you arrived at some of the correct answers, you should review the appropriate chapters. Finally, fill out the Self-Evaluation Chart on page 341 in order to pinpoint the topics that gave you the most difficulty.

50 questions: 1 hour

Directions: Decide which answer choice is best. If the exact numerical value is not one of the answer choices, select the closest approximation. Fill in the circle on the answer sheet that corresponds to your choice.

Notes:
(1) You will need to use a scientific or graphing calculator to answer some of the questions.
(2) You will have to decide whether to put your calculator in degree or radian mode for some problems.
(3) All figures that accompany problems are plane figures unless otherwise stated. Figures are drawn as accurately as possible to provide useful information for solving the problem, except when it is stated in a particular problem that the figure is not drawn to scale.
(4) Unless otherwise indicated, the domain of a function is the set of all real numbers for which the functional value is also a real number.

Reference Information. The following formulas are provided for your information.

Volume of a right circular cone with radius r and height h: $V = \dfrac{1}{3}\pi r^2 h$

Volume of a sphere of radius r: $V = \dfrac{4}{3}\pi r^3$

Surface area of a sphere of radius r: $S = 4\pi r^2$

Volume of a pyramid of base area B and height h: $V = \dfrac{1}{3}Bh$

1. If $f(x) = \dfrac{6x^2 - 3}{2 - 2x^2}$, what value does $f(x)$ approach as x gets infinitely larger?

(A) -3

(B) $-\dfrac{3}{2}$

(C) $\dfrac{3}{2}$

(D) 3

(E) 6

2. Two 6-sided dice are rolled. What is the probability that the sum of the faces showing up is less than 5?

(A) $\dfrac{1}{12}$

(B) $\dfrac{1}{6}$

(C) $\dfrac{7}{11}$

(D) $\dfrac{7}{10}$

(E) $\dfrac{5}{6}$

3. If (a,b) is a solution of the system of equations
$$\begin{cases} 2x - y = 7 \\ x + y = 8 \end{cases}$$, then the difference, $a - b$, equals

(A) -12
(B) -10
(C) 0
(D) 2
(E) 4

4. If $f(x) = x - 1$, $g(x) = 3x$, and $h(x) = \dfrac{5}{x}$, then $f^{-1}(g(h(5))) =$

(A) 4
(B) 2

(C) $\dfrac{5}{6}$

(D) $\dfrac{1}{2}$

(E) $\dfrac{5}{12}$

5. A sphere is inscribed in a cube. The ratio of the volume of the sphere to the volume of the cube is

(A) $0.79 : 1$
(B) $1 : 2$
(C) $0.52 : 1$
(D) $1 : 3.1$
(E) $0.24 : 1$

6. Find y if the slope of the line that contains the points $(-1,3)$ and $(4,y)$ is 0.75.

(A) 0.75
(B) 1
(C) 6.75
(D) 8
(E) 9.67

7. The roots of the equation $3x^4 + 4x^3 + x - 1 = 0$ consist of

(A) three positive real numbers and one negative real number
(B) three negative real numbers and one positive real number
(C) one negative real number and three imaginary numbers
(D) one positive real number, one negative real number, and two imaginary numbers
(E) two positive real numbers, one negative real number, and one imaginary number

8. For what value(s) of k is $x^2 + 3x + k$ divisible by $x + k$?

(A) only 0
(B) only 0 or 2
(C) only 0 or -4
(D) no value of k
(E) any value of k

9. What number should be added to each of the three numbers 3, 11, and 27 so that the resulting numbers form a geometric sequence?

(A) 2
(B) 3
(C) 4
(D) 5
(E) 6

10. What is the equation of the set of points that are 5 units from point (2,3,4)?

(A) $2x + 3y + 4z = 5$
(B) $x^2 + y^2 + z^2 - 4x - 6y - 8z = 25$
(C) $(x - 2)^2 + (y - 3)^2 + (z - 4)^2 = 25$
(D) $x^2 + y^2 + z^2 = 5$
(E) $\dfrac{x}{2} + \dfrac{y}{3} + \dfrac{z}{4} = 5$

11. If $3x^{3/2} = 4$, then $x =$

(A) 1.12
(B) 1.21
(C) 1.34
(D) 1.49
(E) 1.53

12. If $f(x) = x^3 - 4$, then the inverse of $f =$

(A) $-x^3 + 4$

(B) $\sqrt[3]{x + 4}$

(C) $\sqrt[3]{x - 4}$

(D) $\dfrac{1}{x^3 - 4}$

(E) $\dfrac{4}{\sqrt[3]{x}}$

13. If f is an odd function and $f(a) = b$, which of the following must also be true?

  I. $f(a) = -b$
  II. $f(-a) = b$
  III. $f(-a) = -b$

(A) only I
(B) only II
(C) only III
(D) only I and II
(E) only II and III

14. For all θ, $\tan\theta + \cos\theta + \tan(-\theta) + \cos(-\theta) =$

(A) 0
(B) $2\tan\theta$
(C) $2\cos\theta$
(D) $2(\tan\theta + \cos\theta)$
(E) 2

15. The period of the function $f(x) = k\cos kx$ is $\dfrac{\pi}{2}$. The amplitude of f is

(A) $\dfrac{1}{4}$

(B) $\dfrac{1}{2}$

(C) 1
(D) 2
(E) 4

16. If $f(x) = \dfrac{x+2}{(x-2)(x^2-4)}$, its graph will have

 (A) one horizontal and three vertical asymptotes
 (B) one horizontal and two vertical asymptotes
 (C) one horizontal and one vertical asymptote
 (D) zero horizontal and one vertical asymptote
 (E) zero horizontal and two vertical asymptotes

17. At a distance of 100 feet, the angle of elevation from the horizontal ground to the top of a building is 42°. The height of the building is

 (A) 67 feet
 (B) 74 feet
 (C) 90 feet
 (D) 110 feet
 (E) 229 feet

18. A sphere has a surface area of 36π. Its volume is

 (A) 84
 (B) 113
 (C) 201
 (D) 339
 (E) 905

19. A pair of dice is tossed 10 times. What is the probability that no 7s or 11s appear as the sum of the sides facing up?

 (A) 0.08
 (B) 0.09
 (C) 0.11
 (D) 0.16
 (E) 0.24

20. The lengths of two sides of a triangle are 50 inches and 63 inches. The angle opposite the 63-inch side is 66°. How many degrees are in the largest angle of the triangle?

 (A) 66°
 (B) 67°
 (C) 68°
 (D) 71°
 (E) 72°

21. Which of the following is an equation of a line that is perpendicular to $5x + 2y = 8$?

 (A) $8x - 2y = 5$
 (B) $5x - 2y = 8$
 (C) $2x - 5y = 4$
 (D) $2x + 5y = 10$
 (E) $y = \dfrac{2}{-5x+8}$

22. What is the period of the graph of the function

$$y = \frac{\sin x}{1 + \cos x}?$$

(A) 4π

(B) 2π

(C) π

(D) $\dfrac{\pi}{2}$

(E) $\dfrac{\pi}{4}$

23. For what values of k are the roots of the equation $kx^2 + 4x + k = 0$ real and unequal?

(A) $0 < k < 2$

(B) $|k| < 2$

(C) $|k| > 2$

(D) $k > 2$

(E) $-2 < k < 0$ or $0 < k < 2$

24. Minor defects are found on 7 of 10 new cars. If 3 of the 10 cars are selected at random, what is the probability that 2 have minor defects?

(A) 0.143

(B) 0.333

(C) 0.525

(D) 0.667

(E) 0.700

25. If $f(x) = 3x^2 + 24x - 53$, find the negative value of $f^{-1}(0)$.

(A) -58.80

(B) -9.80

(C) -8.23

(D) -1.87

(E) -0.21

26. If $\log_b x = 0.2$ and $\log_b y = 0.4$, what is the relationship between x and y?

(A) $y = 2x$

(B) $y = x^2$

(C) $y = \sqrt{x}$

(D) $xy = 0.6$

(E) $xy = \dfrac{\log 0.6}{\log b}$

27. If $7^{x-1} = 6^x$, find x.

   (A) −13.2
   (B) 0.08
   (C) 0.22
   (D) 0.52
   (E) 12.6

28. A red box contains eight items, of which three are defective, and a blue box contains five items, of which two are defective. An item is drawn at random from each box. What is the probability that one item is defective and one is not?

   (A) $\dfrac{17}{20}$

   (B) $\dfrac{5}{8}$

   (C) $\dfrac{17}{32}$

   (D) $\dfrac{19}{40}$

   (E) $\dfrac{9}{40}$

29. If $(\log_3 x)(\log_5 3) = 3$, find x.

   (A) 5
   (B) 9
   (C) 25
   (D) 81
   (E) 125

30. If $f(x) = \sqrt{x}$, $g(x) = \sqrt[3]{x+1}$, and $h(x) = \sqrt[4]{x+2}$, then $f(g(h(2))) =$

   (A) 1.2
   (B) 1.4
   (C) 2.9
   (D) 4.7
   (E) 8.5

31. In $\triangle ABC$, $\angle A = 45°$, $\angle B = 30°$, and $b = 8$. Side $a =$

   (A) 6.5
   (B) 11
   (C) 12
   (D) 14
   (E) 16

32. The equations of the asymptotes of the graph of $4x^2 - 9y^2 = 36$ are

(A) $y = x$ and $y = -x$

(B) $y = 0$ and $x = 0$

(C) $y = \dfrac{2}{3}x$ and $y = -\dfrac{2}{3}x$

(D) $y = \dfrac{3}{2}x$ and $y = -\dfrac{3}{2}x$

(E) $y = \dfrac{4}{9}x$ and $y = -\dfrac{4}{9}x$

33. If $g(x - 1) = x^2 + 2$, then $g(x) =$

(A) $x^2 - 2x + 3$
(B) $x^2 + 2x + 3$
(C) $x^2 - 3x + 2$
(D) $x^2 + 2$
(E) $x^2 - 2$

34. If $f(x) = 3x^3 - 2x^2 + x - 2$, and $i = \sqrt{-1}$ then $f(i) =$

(A) $-2i - 4$
(B) $4i - 4$
(C) $4i$
(D) $-2i$
(E) 0

35. If the hour hand of a clock moves k radians in 48 minutes, $k =$

(A) 0.3
(B) 0.4
(C) 0.5
(D) 2.4
(E) 5

36. If the longer diagonal of a rhombus is 10 and the larger angle is 100°, what is the area of the rhombus?

(A) 37
(B) 40
(C) 42
(D) 45
(E) 50

37. Let $f(x) = \sqrt{x^3 - 4x}$ and $g(x) = 3x$. The sum of all values of x for which $f(x) = g(x)$ is

(A) −8.5
(B) 0
(C) 8
(D) 9
(E) 9.4

PRACTICE TEST 6

38. How many subsets does a set with n distinct elements have?

(A) n^2

(B) 2^n

(C) $\dfrac{(2n)!}{2(n!)}$

(D) n

(E) $n!$

39. If $f(x) = 2^{3x-5}$, find $f^{-1}(16)$.

(A) 1

(B) 2

(C) 3

(D) 4

(E) 5

40. For what positive value of n are the zeros of $P(x) = 5x^2 + nx + 12$ in ratio $2 : 3$?

(A) 0.42

(B) 1.32

(C) 4.56

(D) 15.8

(E) 25

41. If $f(-x) = -f(x)$ for all x and if the point $(-2,3)$ is on the graph of f, which of the following points must also be on the graph of f?

(A) $(-3,2)$

(B) $(2,-3)$

(C) $(-2,3)$

(D) $(-2,-3)$

(E) $(3,-2)$

42. A man piles 150 toothpicks in layers so that each layer has one less toothpick than the layer below. If the top layer has three toothpicks, how many layers are there?

(A) 15

(B) 17

(C) 20

(D) 148

(E) 11,322

43. If the circle $x^2 + y^2 - 2x - 6y = r^2 - 10$ is tangent to the line $12y = 60$, the value of r is

(A) 1

(B) 2

(C) 3

(D) 4

(E) 5

44. If $a_0 = 0.4$ and $a_{n+1} = 2|a_n| - 1$, then $a_5 =$

(A) -0.6
(B) -0.2
(C) 0.2
(D) 0.4
(E) 0.6

45. If $5.21^p = 2.86^q$, what is the value of $\dfrac{p}{q}$?

(A) -0.60
(B) 0.55
(C) 0.60
(D) 0.64
(E) 1.57

46. The matrix $A = \begin{bmatrix} -3 & x \\ 4 & \dfrac{2}{3} \end{bmatrix}$ does **not** have an inverse. Find the value of x.

(A) $-\dfrac{2}{3}$

(B) $-\dfrac{1}{2}$

(C) 0

(D) $\dfrac{3}{4}$

(E) 1

47. There is a linear relationship between the number of chirps made by a cricket and the air temperature. A least-squares fit of data collected by a biologist yields the equation:

$$\text{temp (°F)} = 22.8 + (3.4)(\#\text{chirps/min})$$

What is the estimated increase in temperature that corresponds to an increase of 5 chirps per minute?

(A) $3.4°F$
(B) $17.0°F$
(C) $22.8°F$
(D) $26.2°F$
(E) $39.8°F$

48. If the length of the diameter of a circle is equal to the length of the major axis of the ellipse whose equation is $x^2 + 4y^2 - 4x + 8y - 28 = 0$, to the nearest whole number, what is the area of the circle?

(A) 28
(B) 64
(C) 113
(D) 254
(E) 452

PRACTICE TEST 6

49. The force of the wind on a sail varies jointly as the area of the sail and the square of the wind velocity. On a sail of area 50 square yards, the force of a 15-mile-per-hour wind is 45 pounds. Find the force on the sail if the wind increases to 45 miles per hour.

(A) 135 pounds
(B) 225 pounds
(C) 405 pounds
(D) 450 pounds
(E) 675 pounds

50. If the riser of each step in the drawing above is 6 inches and the tread is 8 inches, what is the value of $|AB|$?

(A) 40 inches
(B) 43.9 inches
(C) 46.6 inches
(D) 48.3 inches
(E) 50 inches

STOP

If there is still time remaining, you may review your answers.

1. **A**	14. **C**	27. **E**	40. **D**
2. **B**	15. **E**	28. **D**	41. **B**
3. **D**	16. **C**	29. **E**	42. **A**
4. **A**	17. **C**	30. **A**	43. **B**
5. **C**	18. **B**	31. **B**	44. **C**
6. **C**	19. **A**	32. **C**	45. **D**
7. **D**	20. **C**	33. **B**	46. **B**
8. **B**	21. **C**	34. **D**	47. **B**
9. **D**	22. **B**	35. **B**	48. **C**
10. **C**	23. **E**	36. **C**	49. **C**
11. **B**	24. **C**	37. **E**	50. **B**
12. **B**	25. **B**	38. **B**	
13. **C**	26. **B**	39. **C**	

ANSWERS EXPLAINED

The following explanations are keyed to the review portions of this book. The number in brackets after each explanation indicates the appropriate chapter in the Review of Major Topics (Part 2).

An asterisk appears next to those solutions for which a graphing calculator is necessary or helpful.

1. **(A)** The x^2 term "dominates" both the numerator and the denominator, so for very large values of $x, f(x)$ resembles $\dfrac{6x^2}{-2x^2} = -3$. [9]

2. **(B)** There are 36 possible outcomes when two dice are rolled. The possible rolls can be represented by ordered pairs, where the first entry is the outcome of one die and the second is the outcome of the other. Thus, the 36 pairs are (1 and 1), (1 and 2), . . . , (1 and 6), (2 and 1), (2 and 2), . . . , (2 and 6), . . . , (6 and 1), (6 and 2), . . . , (6 and 6). Six pairs have sums less than 5: (1 and 1), (1 and 2), (1 and 3), (2 and 1), (2 and 2), (3 and 1). Therefore, the probability that the sum is less than 5 is $\dfrac{6}{36} = \dfrac{1}{6}$. [22]

3. **(D)** Adding the equations gives $3x = 15$. $x = 5$ and $y = 3$. $a - b = 2$. [2]

4. **(A)** $h(5) = 1$. $g(1) = 3$. Interchange x and y to find that $f^{-1}(x) = x + 1$, and so $f^{-1}(3) = 4$. [1]

* 5. **(C)** Diameter of sphere = side of cube.

    Volume of sphere = $\dfrac{4}{3}\pi r^3$.

    Volume of cube = $s^3 = (2r)^3 = 8r^3$.

    $$\frac{\text{Volume of sphere}}{\text{Volume of cube}} = \frac{\dfrac{4}{3}\pi r^3}{8r^3} = \frac{\pi}{6} \approx \frac{3.14}{6} \approx \frac{0.52}{1} \quad [15]$$

6. **(C)** The desired slope is $\dfrac{y-3}{4+1} = \dfrac{3}{4}$. Cross-multiply and solve for y: $4y - 12 = 15$, so $y = 6.75$. [1]

* 7. **(D)** Plot the graph of $y = 3x^4 + 4x^3 + x - 1$ in the standard window. Observe that the graph crosses the x-axis twice—once at a positive x-value and once at a negative one. Since the function is a degree 4 polynomial, there are 4 roots, so the other two must be complex conjugates. [4]

8. **(B)** If $x^2 + 3x + k$ is divisible by $x + k$, then $(-k)^2 + 3(-k) + k = 0$, or $k^2 - 2k = 0$. Factoring and solving yields the correct answer choice. [4]

9. **(D)** If x is the number, then to have a geometric sequence, $\dfrac{11+x}{3+x} = \dfrac{27+x}{11+x}$.

Cross-multiplying yields $121 + 22x + x^2 = 81 + 30x + x^2$, and subtracting x^2 and solving gives the desired solution.

An alternative solution is to backsolve. Determine which answer choice yields a sequence of 3 integers with a common ratio. [19]

10. **(C)** The set of points represents a sphere with equation $(x - 2)^2 + (y - 3)^2 + (z - 4)^2 = 5^2$. [15]

* 11. **(B)** Plot the graph of $y = 3x^{3/2} - 4$ in the standard window and use CALC/zero to find the correct answer choice.

An alternative solution is to divide the equation by 3 to get $x^{3/2} = \dfrac{4}{3}$ and then raise both sides to the $\dfrac{2}{3}$ power: $x = \left(\dfrac{4}{3}\right)^{2/3} \approx 1.21$. [8]

12. **(B)** Let $y = f(x) = x^3 - 4$. To get the inverse, interchange x and y and solve for y.

$x = y^3 - 4$. $y = \sqrt[3]{x + 4}$. [1]

13. **(C)** Use the definition of an odd function. If $f(a) = b$, then $f(-a) = -b$. Only III is true. [1]

14. **(C)** Tangent is an odd function, so $\tan(-\theta) = -\tan\theta$, and cosine is an even function, so $\cos(-\theta) = \cos\theta$. Therefore, the sum in the problem is $2\cos\theta$. [7]

15. **(E)** Period $= \dfrac{2\pi}{k} = \dfrac{\pi}{2}$. $k = 4$. Amplitude $= 4$. [7]

* 16. **(C)** Plot the graph of $y = \dfrac{x+2}{(x-2)(x^2-4)}$ in the standard window and observe one vertical asymptote ($x = 2$) and one horizontal asymptote ($y = 0$). [9]

* 17. **(C)** Tan $42° = \dfrac{x}{100}$, so $x = 100 \tan 42 \approx 90$. [6]

* 18. **(B)** $4\pi r^2 = 36\pi$. $r^2 = 9$. $r = 3$. $V = \dfrac{4}{3}\pi r^3 = 36\pi \approx 113$. [15]

* 19. **(A)** There are six ways to get a 7 and two ways to get an 11 on two dice, and so there are $36 - 8 = 28$ ways to get anything else. Therefore:

$$P(\text{no 7 or 11}) = P(\text{always getting something else}) = \left(\dfrac{28}{36}\right)^{10} \approx 0.08 \quad [22]$$

* 20. **(C)** Use the Law of Sines: $\dfrac{\sin B}{50} = \dfrac{\sin 66°}{63}$. $\sin B = \dfrac{50\sin 66°}{63} \approx \dfrac{45.68}{63} \approx 0.725$.

$B = \sin^{-1}(0.725) \approx 46°$ and $\angle A = 180° - 46° - 66° = 68°$. [6]

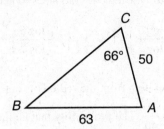

21. **(C)** Solve the equation for y to find the slope of the line is $-\dfrac{5}{2}$. The slope of a

perpendicular line is $\dfrac{2}{5}$. Inspection of the answer choices yields Choice C as the

correct answer. [2]

* 22. **(B)** With your calculator in radian mode, plot the graph of $y = \sin x/(1 + \cos x)$ and observe that the period is 2π.

An alternative solution is to use the identity $\tan\dfrac{x}{2} = \dfrac{\sin x}{1 + \cos x}$ and the fact that the

period of $\tan x$ is π to deduce the period of $\tan\dfrac{x}{2}$ as $\dfrac{\pi}{\dfrac{1}{2}} = 2\pi$. [7]

23. **(E)** $b^2 - 4ac > 0$. $b^2 - 4ac = 16 - 4k^2 > 0$. $4 > k^2$. So $-2 < k < 2$. However, $k \neq 0$ because if $k = 0$, there would no longer be a quadratic equation. [3]

24. **(C)** The denominator of the probability is $_{10}C_3 = 120$, the number of ways 3 cars can be chosen from 10. There are $_7C_2 = 21$ ways of choosing 2 defective cars from 7, and $_3C_1 = 3$ ways of choosing 1 non-defective car from 3. Therefore, there are $(21)(3) = 63$ ways of choosing 2 defective and 1 non-defective cars, and the probability of this is:

$$\dfrac{63}{120} \approx 0.525 \ [22]$$

* 25. **(B)** Find $f^{-1}(0)$ means to find a value of x that makes $3x^2 + 24x - 53 = 0$. Enter the left side of the equation into Y_1, and key 2nd Trace (zero) to determine that the zero is -9.80. [3]

26. **(B)** Since $\log_b x = 0.2$ and $\log_b y = 0.4$, it follows that $x = b^{0.2}$ and $y = b^{0.4}$. Therefore, $x^2 = (b^{0.2})^2 = b^{0.4} = y$. [8]

27. **(E)** Enter $7^{\wedge}(x-1)-6^{\wedge}x$ into Y_1, and generate an Auto table with TblStart $= -14$ and ΔTbl $= 1$. Scan through values of x and observe a change in the sign of Y_1 between $x = 12$ and $x = 13$. Thus, Choice E is the correct answer choice.

An alternative solution is to take $\log_7$ of both sides to get $x - 1 = \log_7 6^x = x \log_7 6 =$

$x \dfrac{\log 6}{\log 7}$. Therefore, $x\left(1 - \dfrac{\log 6}{\log 7}\right) = 1$, so $x = \left(1 - \dfrac{\log 6}{\log 7}\right)^{-1} \approx 12.6$. [8]

28. **(D)** The probability that an item from the red box is defective and an item from the

blue box is good $= \dfrac{3}{8} \cdot \dfrac{3}{5} = \dfrac{9}{40}$. The probability that an item from the red box is good

and that an item from the blue box is defective $= \dfrac{5}{8} \cdot \dfrac{2}{5} = \dfrac{10}{40}$. Since these are mutually

exclusive events, the answer is $\dfrac{9}{40} + \dfrac{10}{40} = \dfrac{19}{40}$. [22]

29. **(E)** By the change-of-base theorem, $\log_3 x = \dfrac{\log_5 x}{\log_5 3}$. Therefore, $\log_5 x = 3$ and $x = 5^3 = 125$. [8]

30. **(A)** Enter f into Y_1; g into Y_2; and h into Y_3. Then return to the home screen and evaluate $Y_1(Y_2(Y_3(2)))$ to get the correct answer choice.

An alternative solution is to evaluate each function, starting with h in turn:

$$\sqrt{\left(\sqrt[3]{\sqrt[4]{2} + 2} + 1\right)} + \sqrt{\left((4)^{\wedge}(1/4) + 1\right)^{\wedge}(1/3)} \approx 1.2 \ [1]$$

31. **(B)** Use the Law of Sines: $\dfrac{\sin 45°}{a} = \dfrac{\sin 30°}{8}$. Therefore $a = \dfrac{8 \sin 45°}{\sin 30°} \approx 11$. [6]

32. **(C)** From this form of the equation of the hyperbola, $\dfrac{x^2}{9} - \dfrac{y^2}{4} = 1$, the equations of

the asymptotes can be found by solving $\dfrac{x^2}{9} - \dfrac{y^2}{4} = 0$. Thus, $y = \pm\dfrac{2}{3}x$. [13]

33. **(B)** Since $x = (x + 1 - 1)$, $g(x) = g((x + 1) - 1) = (x + 1)^2 + 2 = x^2 + 2x + 3$. [12]

34. **(D)** Evaluate $f(i) = 3i^3 - 2i^2 + i - 2 = -2i$ on your graphing calculator. [17]

✳ **35.** **(B)** In 1 hour, the hour hand moves $\frac{1}{12}$ of the way around the clock, or $\frac{2\pi}{12} = \frac{\pi}{6}$ radians. $\frac{48}{60} \cdot \frac{\pi}{6} = \frac{2\pi}{15} \approx \frac{6.28}{15} \approx 0.4.$ [7]

✳ **36.** **(C)** Diagonals of a rhombus are perpendicular; they bisect each other, and they bisect angles of the rhombus. From the figure below, $\tan 40° = \frac{x}{5}$, $x = 5 \tan 40° \approx 4.195$.

$$A = \frac{1}{2}d_1 d_2 = \frac{1}{2}(10)(2)(4.195) \approx 42 \; [6]$$

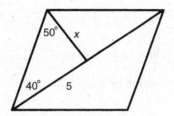

✳ **37.** **(E)** Plot the graphs of f and g in a $[-10,15]$ by $[-10,50]$ window. One point of intersection is the origin. Find the other by using CALC/intersection to get the correct answer choice.

An alternative solution is to set $\sqrt{x^3 - 4x} = 3x$, square both sides, and factor out x to get $x(x^2 - 9x - 4) = 0$. Use the Quadratic Formula with the second factor to find the solutions $x \approx -0.4, 9.4$, and observe that the first does not satisfy the original equation. Thus, the solution is 9.4. [1]

38. **(B)** Each element is either in a subset or not, so there are 2 choices for each of n elements. This yields $2 \times 2 \times \cdots \times 2$ (n factors) $= 2^n$ subsets. [16]

39. **(C)** This problem is most readily solved by substituting each answer choice into f to determine which produces the value 16. Since $f(3) = 16, f^{-1}(16) = 3$. [8]

✳ **40.** **(D)** If the zeros of $P(x)$ are in the ratio $2:3$, they must take the form $2k$ and $3k$ for some value k, and $(x - 2k)(x - 3k) = x^2 - 5k + 6k^2 = 0$. Dividing $P(x)$ by 5 and equating coefficients yields $\frac{n}{5} = -5k$ and $\frac{12}{5} = 6k^2$. Therefore, $k = \pm\sqrt{\frac{2}{5}}$. Since the problem asks for a positive value of n, we use $k = -\sqrt{\frac{2}{5}}$, so $n = -25k \approx 15.8$. [3]

41. **(B)** The function f is odd, so it is symmetrical about the origin. If $(-2,3)$ is on the graph of f, $(2,-3)$ must also be on the graph.

An alternative solution is to conclude $f(-2) = 3$ since $(-2,3)$ is on the graph of f. Therefore, $f(2) = -3$, so the point $(2,-3)$ is also on the graph. [1]

42. **(A)** This is an arithmetic series with $t_1 = 3, d = 1$, and $S = 150$. $150 = \frac{n}{2}[6 + (n-1) \cdot 1]$. $n = 15$. [19]

43. **(B)** Complete the square in the equation of the circle: $(x - 1)^2 + (y - 3)^2 = r^2$. The center of the circle is at (1,3). If the circle is tangent to the line, then the distance between (1,3) and the line $y = 5$ is the radius of the circle, r. Therefore, $r = 2$. [15]

* 44. **(C)** Enter 0.4 into your calculator, followed by $2|Ans| - 1$ five times to get $a_5 = 0.2$.

An alternative solution is to evaluate each a_i in turn:

$$a_1 = 2|0.4| - 1 = -0.2$$
$$a_2 = 2|a_1| - 1 = 2|-0.2| - 1 = -0.6$$
$$a_3 = 2|a_2| - 1 = 2|-0.6| - 1 = 0.2$$
$$a_4 = 2|a_3| - 1 = 2|0.2| - 1 = -0.6$$
$$a_5 = 2|a_4| - 1 = 2|-0.6| - 1 = 0.2 \text{ [19]}$$

* 45. **(D)** Take the logarithms (either base 10 or base e) to get $p \log 5.21 = q \log 2.86$.

Divide both sides of this equation by $q \log 5.21$ to get $\dfrac{p}{q} = \dfrac{\log 2.86}{\log 5.21} \approx 0.64$. [8]

46. **(B)** Since matrix A doesn't have an inverse, $\det A = |A| = 0$. Therefore:

$$(-3)\left(\frac{2}{3}\right) - 4x = 0, \text{ so } x = -\frac{1}{2} \text{ [18]}$$

* 47. **(B)** The temperature increases by 3.4°F for each additional chirp. Therefore, 5 additional chirps indicate an increase of $5(3.4) = 17.0$°F. [21]

* 48. **(C)** Complete the square on the ellipse formula, and put the equation in standard form:

$x^2 - 4x + 4 + 4(y^2 + 2y + 1) = 28 + 4 + 4.$ $\dfrac{(x-2)^2}{36} + \dfrac{(y+1)^2}{9} = 1.$ This leads to the

length of the major axis: $2\sqrt{36} = 12$. Therefore, the radius of the circle is 6, and the area is $36\pi \approx 113$. [13]

49. **(C)** Since the velocity of a 45-mile-per-hour wind is 3 times that of a 15-mile-per-hour wind and the force on the sail is proportional to the square of the wind velocity, the force on the sail of a 45-mile-per-hour wind is 9 times that of a 15-mile-per-hour wind:

$$9 \cdot 45 = 405 \text{ [1]}$$

* 50. **(B)** Total horizontal distance traveled is $(4)(8) = 32$. Total vertical distance traveled is $(5)(6) = 30$. If a coordinate system is superimposed on the diagram with A at (0,0), then B is at (32,30). Use the distance formula to find the distance between two points to

compute the correct answer choice: $\sqrt{32^2 + 30^2} \approx 43.9$. [2]

Self-Evaluation Chart for Practice Test 6

Subject Area	Questions and Review Section						Right	Number Wrong	Omitted
Algebra and Functions (24 questions)	1 9	3 2	4 1	6 1	7 4	8 4	_____	_____	_____
	11 8	12 1	13 1	16 9	21 2	23 3	_____	_____	_____
	25 3	26 8	27 8	29 8	30 1	37 1	_____	_____	_____
	39 8	40 3	41 1	45 8	49 1	50 2	_____	_____	_____
Trigonometry (8 questions)	14 7	15 7	17 6	20 6	22 7	31 6	_____	_____	_____
	35 7	36 6					_____	_____	_____
Coordinate and Three-Dimensional Geometry (6 questions)	5 15	10 15	18 15	32 13	43 15	48 13	_____	_____	_____
Number and Operations (7 questions)	9 19	33 12	34 17	38 16	42 19	44 19	_____	_____	_____
	46 18						_____	_____	_____
Data Analysis, Statistics, and Probability (5 questions)	2 22	19 22	24 22	28 22	47 21		_____	_____	_____
TOTALS							_____	_____	_____

Evaluate Your Performance on Practice Test 6

Rating	Number Right
Excellent	41–50
Very good	33–40
Above average	25–32
Average	15–24
Below average	Below 15

CALCULATING YOUR SCORE

Raw score R = number right $- \frac{1}{4}$ (number wrong), rounded = _____

Approximate scaled score $S = 800 - 10(44 - R) =$ _____

If $R \geq 44$, $S = 800$.

Summary of Formulas

CHAPTER 2: LINEAR FUNCTIONS

General equation of a line: $Ax + By + C = 0$

Slope-intercept form: $y = mx + b$, where m represents the slope and b represents the y-intercept

Point-slope form: $y - y_1 = m(x - x_1)$, where m represents the slope and (x_1, y_1) are the coordinates of some point on the line

Slope: $m = \dfrac{y_2 - y_1}{x_2 - x_1}$, where (x_1, y_1) and (x_2, y_2) are the coordinates of two points

Parallel lines have equal slopes.

Perpendicular lines have slopes that are negative reciprocals.

If m_1 and m_2 are the slopes of two perpendicular lines, $m_1 \cdot m_2 = -1$.

Distance between two points with coordinates (x_1, y_1) and (x_2, y_2): $\sqrt{\left(x_2 - x_1\right)^2 + \left(y_2 - y_1\right)^2}$

Coordinates of the midpoint between two points:

$$\left(\frac{x_1 + x_2}{2}, \frac{y_1 + y_2}{2} \right)$$

Distance between a point with coordinates (x_1, y_1) and a line $Ax + By + C = 0$: $\dfrac{\left| Ax_1 + By_1 + C \right|}{\sqrt{A^2 + B^2}}$

If θ is the angle between two lines, $\tan\theta = \dfrac{m_1 - m_2}{1 + m_1 m_2}$, where m_1 and m_2 are the slopes of the two lines.

CHAPTER 3: QUADRATIC FUNCTIONS

General quadratic equation: $ax^2 + bx + c = 0$

General quadratic formula: $x = \dfrac{-b \pm \sqrt{b^2 - 4ac}}{2a}$

General quadratic function: $y = ax^2 + bx + c$

Coordinates of vertex: $\left(-\dfrac{b}{2a}, c - \dfrac{b^2}{4a} \right)$

Axis of symmetry equation: $x = -\dfrac{b}{2a}$

Sum of zeros (roots) $= -\dfrac{b}{a}$

Product of zeros (roots) $= \dfrac{c}{a}$

Nature of zeros (roots):

If $b^2 - 4ac < 0$, two complex numbers

If $b^2 - 4ac = 0$, two equal real numbers

If $b^2 - 4ac > 0$, two unequal real numbers

CHAPTER 6: TRIANGLE TRIGONOMETRY

$\sin\theta = \dfrac{\text{opposite}}{\text{hypotenuse}}$ $\cos\theta = \dfrac{\text{adjacent}}{\text{hypotenuse}}$

$\tan\theta = \dfrac{\text{opposite}}{\text{adjacent}}$ $\cot\theta = \dfrac{\text{adjacent}}{\text{opposite}}$

$\sec\theta = \dfrac{\text{hypotenuse}}{\text{adjacent}}$ $\csc\theta = \dfrac{\text{hypotenuse}}{\text{opposite}}$

In any $\triangle ABC$:

Law of Sines: $\dfrac{\sin A}{a} = \dfrac{\sin B}{b} = \dfrac{\sin C}{c}$

Law of Cosines: $a^2 = b^2 + c^2 - 2bc \cos A$

$\quad\quad\quad\quad\quad\quad b^2 = a^2 + c^2 - 2ac \cos B$

$\quad\quad\quad\quad\quad\quad c^2 = a^2 + b^2 - 2ab \cos C$

Area $= \dfrac{1}{2} bc \cdot \sin A$

CHAPTER 7: TRIGONOMETRIC FUNCTIONS

$$\pi^R = 180°$$

Length of arc in circle of radius r and central angle θ is given by $r\theta$. (θ in radians)

Area of sector of circle of radius r and central angle θ is given by $\dfrac{1}{2} r^2 \theta$. (θ in radians)

1. $\sin^2 x + \cos^2 x = 1$ ⎤
2. $\tan^2 x + 1 = \sec^2 x$ ⎬ Pythagorean identities
3. $\cot^2 x + 1 = \csc^2 x$ ⎦

4. $\sin 2A = 2\sin A \cos A$ ⎤
5. $\cos 2A = \cos^2 A - \sin^2 A$ ⎬
6. $\quad\quad\quad = 2\cos^2 A - 1$ ⎬ double-angle
7. $\quad\quad\quad = 1 - 2\sin^2 A$ ⎬ formulas
8. $\tan 2A = \dfrac{2\tan A}{1 - \tan^2 A}$ ⎦

CHAPTER 8: EXPONENTIAL AND LOGARITHMIC FUNCTIONS

Exponents

$x^a \cdot x^b = x^{a+b} \quad\quad \dfrac{x^a}{x^b} = x^{a-b}$

$(x^a)^b = x^{ab}$

$x^0 = 1 \quad\quad x^{-a} = \dfrac{1}{x^a}$

Logarithms

$\log_b(pq) = \log_b p + \log_b q$

$\log_b p^x = x \log_b p$

$\log_b p = \dfrac{\log_a p}{\log_a b}$

$\log_b\left(\dfrac{p}{q}\right) = \log_b p - \log_b q$

$\log_b 1 = 0$

$\log_b b = 1$

$b^{\log_b P} = p$

$\text{Log}_b N = x$ if and only if $b^x = N$

CHAPTER 11: PIECEWISE FUNCTIONS

Absolute Value

If $x \geq 0$, then $|x| = x$.

If $x < 0$, then $|x| = -x$.

Greatest Integer Function

$[x] = i$, where i is an integer and $i \leq x < i + 1$

CHAPTER 13: CONIC SECTIONS

Standard Equation of a Circle

$x^2 + y^2 = r^2$

with radius $= r$

Standard Equation of an Ellipse

$\dfrac{x^2}{a^2} + \dfrac{y^2}{b^2} = 1$, major axis horizontal

$\dfrac{x^2}{b^2} + \dfrac{y^2}{a^2} = 1$, major axis vertical,

where $a^2 = b^2 + c^2$

Standard Equation of a Hyperbola

$\dfrac{x^2}{a^2} - \dfrac{y^2}{b^2} = 1$, transverse axis horizontal

$\dfrac{y^2}{a^2} - \dfrac{x^2}{b^2} = 1$, transverse axis vertical,

where $c^2 = a^2 + b^2$

CHAPTER 14: POLAR COORDINATES

$x = r \cos \theta$ $y = r \sin \theta$

$x^2 + y^2 = r^2$

CHAPTER 15: THREE-DIMENSIONAL GEOMETRY

Distance between two points with coordinates

(x_1, y_1, z_1) and $(x_2, y_2, z_2) =$

$$\sqrt{\left(x_2 - x_1\right)^2 + \left(y_2 - y_1\right)^2 + \left(z_2 - z_1\right)^2}$$

Distance between a point with coordinates (x_1, y_1, z_1) and a plane with equation

$$Ax + By + Cz + D = 0 = \frac{Ax_1 + By_1 + Cz_1 + D}{\sqrt{A^2 + B^2 + C^2}}$$

Cylinder Formulas

Volume $= \pi r^2 h$

Lateral surface area $= 2\pi rh$

Total surface area $= 2\pi rh + 2\pi r^2$

In all formulas, r = radius of base, h = height

Cone Formulas

Volume $= \dfrac{1}{3}\pi r^2 h$

Total surface area $= \pi r\sqrt{r^2 + h^2} + \pi r^2$

In all formulas, r = radius of base, h = height

Sphere Formulas

Volume $= \dfrac{4}{3}\pi r^3$

Surface area $= 4\pi r^2$

In both formulas, r = radius

CHAPTER 16: COUNTING

Permutations

$$_nP_r = \frac{n!}{(n-r)!}, \text{ where } n! = n(n-1)(n-2) \cdots 3 \cdot 2 \cdot 1$$

Combinations

$$_nC_r = \begin{pmatrix} n \\ r \end{pmatrix} = \frac{n!}{(n-r)!\,r!} = \frac{_nP_r}{r!}$$

CHAPTER 17: COMPLEX NUMBERS

$i^0 = 1, i^1 = i, i^{-2} = -1, i^3 = -i, i^4 = 1, \ldots$

modulus $|a + bi| = a^2 + b^2$

CHAPTER 18: MATRICES

Determinant of a 2 × 2 Matrix

$$\begin{vmatrix} a & b \\ c & d \end{vmatrix} = ad - bc$$

CHAPTER 19: SEQUENCES AND SERIES

Arithmetic Sequence (or Progression)

nth term: $t_n = t_1 + (n-1)d$

Sum of n terms: $S_n = \dfrac{n}{2}(t_1 + t_n)$

$$= \frac{n}{2}[2t_1 + (n-1)d]$$

Geometric Sequence (or Progression)

nth term: $t_n = t_1 r^{n-1}$

Sum of n terms: $S_n = \dfrac{t_1(1 - r^n)}{1 - r}$

If $|r| < 1$, $S_\infty = \lim\limits_{n \to \infty} S_n = \dfrac{t_1}{1 - r}$

CHAPTER 20: VECTORS

If $\vec{V} = (v_1, v_2)$ and $\vec{U} = (u_1, u_2)$,

$\quad \vec{V} + \vec{U} = (v_1 + u_1, v_2 + u_2)$

$\vec{V} \cdot \vec{U} = v_1 u_1 + v_2 u_2$

Two vectors are perpendicular if and only if $\vec{V} \cdot \vec{U} = 0$.

norm $\|V\| = \sqrt{a^2 + b^2}$ where $V = (a, b)$

CHAPTER 21: STATISTICS

Mean $\bar{x}$ of $x_1, x_2, \ldots, x_n$: $\bar{x} = \dfrac{1}{n} \displaystyle\sum_{i=1}^{n} x_i$

Standard deviation s_x of $x_1, x_2, \ldots, x_n$:

$$s_x = \sqrt{\frac{1}{n} \sum_{i=1}^{n} \left(x_i - \bar{x}\right)^2}$$

CHAPTER 22: PROBABILITY

$P(\text{event}) = \dfrac{\text{number of ways to get a successful result}}{\text{total number of ways of getting any result}}$

Independent events: $P(A \cap B) = P(A) \cdot P(B)$

Mutually exclusive events: $P(A \cap B) = 0$ and
$P(A \cup B) = P(A) + P(B)$

Index

Cylinder
  formula of a, 345
  lateral surface area of, 345
  surface area of, 149, 345
  volume of, 149, 345

D

Data, quantitative, 187
Degenerate conics, 133
Dependent variable, 189
Determinant, 167–168
Determinants of 2×2 matrix, 345
Difference function, 34
Directrix, 134, 138, 141
Discontinuity, 109
Discriminant, 52
Disjunction, 196
Distance between a point with coordinates and a
      plane, 345
Distance between two points, 45–46, 133, 151, 343, 345
Domain. *See also* function
  about, 33, 35, 38, 90
  of all real numbers, 9, 33, 86
  of cube root function, 271
  of function, 33, 315
  of inverse function, 35
  of $\log_{10}$, 268
  of natural numbers, 175
  of piecewise functions, 121
  of quadratic function, 51
  of square root function, 315
  of trig functions, 86, 90
Dot product, 183–184
Double angle formula, 89, 295

E

Ellipse, 33, 116, 133–135, 138, 142–143, 147
End behavior, 57, 60
Entries, 165–166
Equations
  parametric, 115–116
  quadratic, 52, 168, 189, 343
  solving systems of, 168
Even function, 36, 87, 130
Events
  certain, 195
  definition of, 195
  impossible, 195
  independent, 195–196, 346
  mutually exclusive, 196, 346
  probability of, 195

Exponential functions, 103–105, 344
Exponential model, 189
Exponential scatter plot, 189

F

Factor theorem, 59, 274
Factorial, 155–156
Factoring, 59, 168
Favorable outcomes, 195
Fibonacci sequence, 176
Finite sequence, 175
Foci, 134
Focus, 134
Formula
  cone, 345
  coordinate geometry, 120
  cylinder, 345
  double-angle, 344
  exponential function, 103–105
  probability, 346
  quadratic, 51–53, 59–60, 343
  sphere, 345
  trigonometric functions, 344
Frequency, 86
Function(s). *See also* domain; trigonometric function
  about, 33–38
  combining, 33
  composition of, 34
  definition of, 33
  difference, 34
  even, 36, 87, 130
  exponential, 103–105
  greatest integer, 121–122, 344
  inverse of, 34–35
  inverse trig, 86, 90–91
  linear, 45–46, 343
  logarithmic, 103–105, 344
  odd, 36, 86, 130
  parent trig, 86
  period of, 86
  periodic, 86, 88
  piecewise, 38, 119–122, 344
  product, 34
  quadratic, 51–53, 343
  quotient, 34
  rational, 109–111
  standard position of a trigonometric, 81
  step, 121
  sum, 34
  zeros of, 51, 246, 291
Fundamental Theorem of Algebra, 58

Polynomials
    higher degree, 57–60
    linear functions, 45–46
Postage stamp function, 119
Powers of *i*, 161, 245
Prism, 150
    surface area of, 149
    volume of, 149
Probability, 195–197, 346
Product function, 34
Product of an imaginary number, 162
Product of zeros (roots), 343
Pyramid, 149
    surface area of, 149
    volume of, 9, 149, 207, 231, 255, 279, 301, 323
Pythagorean identities, 82, 89, 91, 344
Pythagorean Theorem, 69, 72, 82, 145, 149, 151, 162

Q

Quadrantal angle, 68, 78
Quadratic equation, 52, 168, 189, 343
Quadratic form, 57, 59
Quadratic formula, 51–53, 59–60, 343
Quadratic functions, 51–53, 343
Quadratic model, 189
Quadratic polynomials, 57
Quadratic scatterplot, 189
Quantitative data, 187
Quotient function, 34

R

Radian, 84–85
Radian measure, 84, 90
Radian mode, 91
Range, 33, 35, 38, 51–52
Rational function, 109–112
Rational root theorem, 59
Real numbers, 33, 35, 51, 86, 103, 161, 183, 343
Reciprocal identities, 89
Reciprocals of complex numbers, 161
Rectangular
    array of numbers, 165
    coordinate system, 81
    coordinates, 145–146, 149, 162
Rectangular solid, 149
    surface area of, 149
    volume of, 149
Recursive formula, 176
Recursively defined sequence, 223
Reduced row echelon form, 168
Reference angle, 67–68, 82, 91

Reflection, 35, 121, 127–130, 132, 162
Regression, 189–190, 192
Relation, 33
Remainder theorem, 271
Resultant, 183–184
Rhombus, 330, 339
Right angle, 67, 69, 72
Right circular cone
    about, 150
    volume of, 9, 207, 231, 255, 279, 301, 323
Right prism, 150
Right triangles, 67, 69, 73, 82
Row matrix, 165

S

Sample space, 195
Scalar multiplication, 165
Scatterplot, 189–190
Sec θ, 81–82
Sec^{-1}, 90
Secant, 84, 86, 92
Sequence, 175–179
    arithmetic, 175–177, 345
    finite, 175
    geometric, 178–179
    infinite, 175
    recursively defined, 223
Series, 175–179
Sin θ, 81–82
Sin^{-1}, 90
Singular matrix, 168
Slope
    about, 45–46
    of horizontal line, 45
    of perpendicular lines, 23, 220
Slope-intercept, 45, 343
Special angle, 67
Sphere, 149–151
    formula, 345
    surface area of, 9, 149, 207, 231, 255, 279, 301, 323, 345
    volume of, 9, 149, 207, 231, 255, 279, 301, 323, 345
Square matrix, 165, 167–168
Square root, 33, 53
    of negative number, 161
    of positive number, 161
Standard deviation, 188–189, 346
Standard equation
    of circle, 344
    of conics, 133

of ellipse, 134, 142, 344
of hyperbola, 135, 143, 344
Standard position, 81
  angle, 68, 77
  radian, 77
  of rectangular coordinate system, 81
  of trigonometric function, 81
Statistical
  mean, 187–188
  median, 187–188
  mode, 187
Statistics, 187–190, 346
Step functions, 121
Stretches/shrinks, 86, 127–128
Study plans, 4
Sum
  function, 34
  of n terms, 345
  of zeros (roots), 343
Surface area, 149, 345
  of cone, 149, 345
  of cube, 149
  of cylinder, 149, 345
  of prism, 149
  of pyramid, 149
  of rectangular solid, 149
  of sphere, 9, 149, 207, 231, 255, 279, 301, 323, 345
Symmetry, 127–130
Synthetic division, 59
Systems of equations, 168

T

Tan θ, 81–82
Tan^{-1}, 90
Tangent, 67, 82, 84, 86, 90
Test scoring, 3–4
Three-dimensional geometry, 149–152, 345
Total surface area, 345
Transformation, 127–130
  coordinate geometry, 120
  horizontal, 129
  vertical, 127–130
Translation, 127, 129, 133
  horizontal, 86
  vertical, 86, 88
Transverse axis, 135, 143, 344

Transverse axis horizontal, 344
Transverse axis vertical, 344
Triangle
  ambiguous case, 71, 78
  obtuse, 72–73
  right, 67, 69, 73, 82
  trigonometry, 67–74, 343–344
Trigonometric function, 81–92, 344
  domain of, 86, 90
  graph of, 86–88, 90
  inverse, 86, 90–91
  parent, 86
  range of, 86, 90
  standard position, 81
Trigonometric inequality, 85, 90
Trigonometry, triangle, 67–74, 343–344

V

Vectors, 183–184, 346
Venn diagram, 155, 159
Vertex, 51
Vertex form, 51
Vertical
  transformations, 127–130
  translations, 86–88
Vertical Line Test, 33, 35
Volume
  of cone, 149, 345
  of cube, 149
  of cylinder, 149, 345
  of prism, 149
  of pyramid, 9, 149, 207, 231, 255, 279, 301, 323
  of rectangular solid, 149
  of right circular cone, 9, 207, 231, 255, 279, 301, 323
  of sphere, 9, 149, 207, 231, 255, 279, 301, 323, 345

W

Weighted average, 188

Z

Zero Product Property, 52
Zeros
  of the function, 51, 246, 291
  nature of, 343
  product of, 343
  sum of, 343

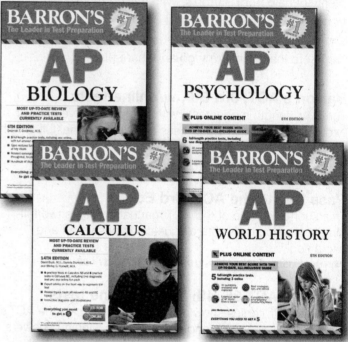